Damned If We Don't!
IDEAS FOR ACCELERATING CHANGE AROUND WATER

Christopher J. Peacock

Water Anthology Press
Granite Bay, CA

To my fellow innovators, collaborators and instigators
who continue pushing the boundaries of change.

Water Anthology Press

Copyright © 2014 by Christopher Peacock
Cover art by Angelina Gonzalez (www.nthreeq.com)
With editing by Craig Martin (www.CMBW.us)

All chapters are copyrighted to their respective authors, and used here with their permission. An extension of this copyright page can be found at the end of this book.

Damned If We Don't!: Ideas for Accelerating Change Around Water / Christopher J. Peacock. —1st ed.
ISBN-13: 978-0692301982
ISBN-10: 0692301984

Water Anthology Press
www.WaterAnthology.com
An imprint of
The Water Innovation Project, LLC

Ordering Information:
Quantity sales. Special discounts are available on quantity purchases by corporations, associations, and others. For details, contact us at the website above.

Jon —

Thank you for your ongoing support, friendship & industry endeavors. May the WolfPack continue to grow & transform the world of water!

No portion of this book may be reproduced by any means, graphic, mechanical, electronic, or otherwise, without first obtaining the permission of the copyright holder except in the case of brief quotations embodied in critical articles or reviews.

Because of the dynamic nature of the Internet, any Web addresses or links contained in this book may have changed since publication and may no longer be valid. The views expressed in this work are solely those of each author, independent of each other.

For more information, contact Water Anthology Press at www.wateranthology.com.

Contents

About ix

Preface x

Introduction (to the conspiracy) 1

Extreme Weather in Canada: Math & Aftermath – R.W. Sandford 3

Integration and Consciousness – Suzanne King 13

The End of Traditional Conferences: How water discussions must evolve – Karen Kun 23

Developing Leadership in the Water Sector – Renee Kayal 31

When Women's Networking Gets Personal (in the good way) – Leila Munla 39

Water Issues are People Issues – Julia Ko 45

Advancing One Water Management with One Water Communication – Donna Vincent Roa, PhD, ABC 49

Utilize Social Media – don't block it. – Todd Danielson, P.E., BCEE 59

Getting beyond dots on a map: Collaborative science and scenario planning in the Upper Gila River Watershed - Kelly Mott Lacroix & Dr. Sharon Megdal 67

Water Clusters and the Acceleration of Innovation – Jon Grant 77

Accelerating Innovation – McGee Young 87

Technology, Utilities and Innovation - from "What if?" to "Here's how" – Jim Ginley..97

Smart Homes, Smarter Consumers – Eric Elias 105

Why We Need to Unlock the Enormous Power of Data in the Water Industry - Sunit Mohindroo & Ahmed Badruddin......... 111

Industrial Ecosystems: 21st century challenges, 21st century solutions – Jim Lauria..123

Convergence: Utilities need to mind the data as well as the drops – Graham Symmonds..133

The Hitchhiker's Guide to Digital Efficiency – Alan Hinchman 141

Non-Revenue Water: A prime target of water innovation – David McGimpsey 149

Tapping into the Power of Behavioral Science: Insights & Opportunities for Water-Use Efficiency – Jeff Lipton & Caroline Hodge 157

Conserve2Enhance: Bridging voluntary water conservation and community-driven environmental enhancement – Candice Rupprecht, Dr. Sharon B. Megdal, Brittany Xiu and Kelly Mott Lacroix 173

Rainfrastructure: Building stormwater utilities – Kevin Mercer 183

From Serious Issues to Serious Games - and Back! – Chengzi Chew & Gareth James Lloyd 191

Acknowledgements 201

About the Editor 203

Copyright Acknowledgements 204

Continue the Discussion 207

About

Damned If We Don't is a series of essays written for members of the water sector *by* members of the water sector.

This book is a microcosm of the potential for collaboration and innovation that already exists in the sector today – it represents a shared vision to help accelerate much needed change around water, whether it's through sharing ideas or encouraging more dialogue about critical issues.

Written by thought leaders who are directly addressing and tackling many of the issues surrounding water, this collection of material shares innovative approaches and provides ideas for the management of our most important resource.

From policy ideas to behavioral change, from engineering to new technology, these authors include solutions and reflect on the importance of collaboration, teamwork and dialogue. The goal is to help people reflect on these growing opportunities that can transform the way we manage water throughout the supply chain.

The water sector is fractured, but we can bring it together through effective dialogue and sharing ideas; *Damned If We Don't* is a step in that direction.

Preface

Promoting change in any industry can be treacherous.

The water sector is no different.

When combined, discussions around changing how we manage water often leads to contentious debates, indecision and all too lengthy periods of inaction. The worst case scenario could result in outright war.

As a professional in the water sector, I have become distraught by our inability to cut across silos – real or imagined – and our failure to find effective levers for change. For too many years, I've watched as we've continually convinced ourselves that water is cheap, if not free.

After witnessing our inability to effectively communicate with the general public on the real issues surrounding water, seeing how the actual impact climate change is having on our water supplies, and the price we should be paying for maintaining a livelihood we all too often take for granted, I knew I couldn't sit around any longer – I needed to take action.

This book is but one of many projects to address the way we manage our most precious resource. What started out as an idea to share some cool ideas from amazing people in the water sector has evolved into a full-fledged publication, which has personally taken me on a fantastic journey.

Writing is not an easy task for most. I am privileged to have met all of the authors contributing to this book. Most of those meetings have involved intense dialogue and a shared passion for accelerating change around the way we manage water. The majority of the authors in this book, with a few notable exceptions, are not prolific authors. In fact, for many of the individuals who have contributed to this work, this is their first "prime time" publication. And, to all of them, a hearty congratulations is in order.

Taking a position, writing about it and actually finding the will power to share it with the world in permanent form is to be commended. There will be some critical reviews, praise, questions and discussions as a result of the works in this anthology – quite honestly, that's the entire point.

So we anticipate pushback, we encourage questions and most importantly, we seek a deeper dialogue that can lead us to sustainable practices. We are sharing our individual ideas and actions taken to not only effect change around water, but accelerate change entirely – so embrace the change!

Christopher J. Peacock
Granite Bay, CA
September 2014

Introduction (to the conspiracy)

During the spring of 2014, as I was getting my arms around this book – realizing what I had actually gotten myself into – I gave a presentation around the concept of Accelerating Change. I laid out a five stage process for the audience to enable their own acceleration:

1. Know your why
2. Understand your value
3. Find the inflection point
4. Build your tribe
5. Do something

As I started receiving completed chapters for this book, I noticed that these stages were embedded in each of the author's chapters and I realized that a conspiracy was now at play. There was trepidation on my part to go through with this publication: there are numerous books, guides and journals that can provide deeper scientific, political, economic and social understandings of the environment for which we have included in this anthology.

I was concerned that this book might be too "soft", "warm" or "fuzzy" for the intended audience; but then again, that was part of the intent.

I had to remind myself that this is a book born of passion and I wanted contributions by people who are truly making a difference by actively doing something – there's already enough talking and not enough doing.

And that is the true conspiracy of this project, aside from getting people thinking about new ideas and engaging them in deeper dialogue: the real purpose for this book is *to become a catalyst and mobilize a revolution around water*. To give people the consent to go and make a difference, the courage to take a step forward, to explore a new idea, to innovate, to challenge, to learn, to grow and to ultimately do something.

This book represents the actions of people who are doing things on a daily basis to accelerate change. It's far from soft, warm or fuzzy; each author has found a way to accelerate change, and I guarantee that they have been challenged by their individual paths along the way. They have made financial commitments to start companies, emotional commitments to follow through on their visions and public commitments to do something.

It doesn't get much more hard-core than putting yourself on the line for an idea, business or research project that may fail. We hope you learn from our ideas or become inspired by our actions.

[1]

Extreme Weather in Canada: Math & Aftermath – R.W. Sandford

WHILE WE MAY NOT HAVE PERCEIVED IT AS SUCH, what is happening hydrologically in the Arctic, and in the north, is exactly what is happening in much of the rest of Canada. Warming is causing the post-glacial hydrological wealth of Canada to change form. Sea ice is vanishing. Snow cover is diminishing in both extent and duration. Some 300 glaciers have disappeared from the Canadian Rockies.

However, the water is not disappearing; water doesn't do that.

What is happening is liquid water moves to a different place in the hydrosphere where it may not be available for our use when we want it – as in the case of extreme drought – and it moves where (under certain circumstances) the water can also cause a lot of damage through flooding.

In 2011, when flooding hit Manitoba, it appeared that this was evidence that the Central Great Plains were approaching – or perhaps had even passed over an invisible threshold – into a new hydro-climatic state. We were seeing things that we hadn't seen or at least fully understood before.

While certain conditions or events have probably existed for an eternity, we have recently discovered, for example, the presence of atmospheric rivers. These corridors of intense winds and moist air can be 400-500 kilometres across, thousands of kilometres long, and they can carry the equivalent of 10 times the average daily discharge of the St. Lawrence River.

Atmospheric rivers produce flooding of the magnitude we saw in Australia and Pakistan in 2010, and it possibly caused the flooding in parts of the Central Great Plains region of North America in 2011. An atmospheric river also broke rainfall records by 87% in the Kootenay-Columbia River of BC in 2012.

You can bet that *all of this* was very much on our minds when, on the evening of Wednesday, June 19th, 2013, it began to rain and forecasters were estimated nearly 150 millimetres would fall. From my viewpoint, this was critical because the rain was falling in Canmore, the town in which I live.

The Alberta Flood of 2013

I would like to acknowledge that we are fortunate to have a minute-by-minute, hour-by-hour chronicle of the Alberta flood at least in one area in which it originated. Dr. John Pomeroy and members of his University of Saskatchewan research team happened to be in Canmore, and they were in the Kananaskis region of Alberta when the flooding began. His observations, and the observations of his research team, form the foundation for this analysis of the math and aftermath of what turned into the largest and most costly flood disaster in the history of Canada.

It is interesting to note that much of the math related to this extreme weather event had already been done well in advance of the flooding. In a report prepared for Environment Canada in 1996, Dr. Pomeroy warned that changes in hydrology would likely increase the potential for flooding particularly on the Canadian prairies. In that report, he observed that fresh water is both a mediator and a transmitter of climate change effects.

He wrote that water should be viewed not just as a substance but as a flow of mass, energy and biochemical constituents through and between ecosystems as well as between the land surface and the atmosphere. Liquid water, water vapour, snow and ice, he said, transmit climate change impacts across the country and across ecological and political boundaries. That appears to be exactly what is happening.

On July 8th, 2013, three weeks after the Alberta disaster, the downtown core of the city of Toronto was devastated by flooding. The math and aftermath of this disaster are straightforward. The uncertainty of a changing climate clearly demands a faster response, but making decisions and funding action in a municipality like Toronto is akin to moving mountains. The crippling cost of replacing a stormwater system (designed for an earlier climatic regime) and a host of competing political interests prevent Toronto from becoming a resilient city.

Unfortunately, the flooding in Alberta and Ontario in June and July of 2013 is nothing compared to what the atmosphere is capable of delivering in the future. This was made evident by a series of extreme weather events that happened in Russia shortly after the flooding in Canada.

The Unexpected Siberian Flood

What happened in Russia a month after the flooding in Alberta is almost beyond imagination. In this instance, the math and aftermath don't add up.

The weakening of the European jet stream, caused by reduced snow and sea ice cover, led to the creation of a heat dome in Northern Siberia. In July, hundreds of wildfires broke out, and they were so hot that they melted the permafrost below the burning forests. This allowed methane gases to be released from the thawing tundra, and that added fuel to the fires.

Then, in the midst of what was coming to resemble a virtual fire storm, three atmospheric rivers collided over the region in early August. The result? Within four days, this created a flood that covered a million square kilometres.

Again we see evidence of a new trend: floods and fires of a magnitude seldom experienced before in the same basin in the same year. But we also see something much darker – it appears that the frozen lid of the jar containing most of the world's methane is melting.

Something Strange is Going On...

At an international scientific conference held in Canmore in the wake of the 2013 flood, it was demonstrated that many of our recent floods were similar – each involved rotating low pressure systems that remained in the same place for an unusual period of time, which brought continuous precipitation up from the south and resulted in long duration, heavy rainfall events that covered very large areas.

There is also concern, however, that the kinds of storms we will have in the future may be fundamentally different in character than what we are used to experiencing. A major flood further south in Colorado, which also occurred in 2013, was different in that it occurred in September. Researchers are also examining other anomalies.

The behaviour of the Colorado storm suggests that its precipitation may have been generated by processes of rain-drop formation, which is more typical of the tropical region where the storm originated instead of from local temperate conditions. The State Climatologist of Colorado, Nolan Doesken, noted that the storm "shattered all records for the most water vapour in the atmosphere." From this, we might surmise that the floods of 2013 offer us a glimpse into the wild weather we might expect in a warmer world.

> So what is going on out there? Something certainly appears amiss. You don't need to be a scientist to figure out that our weather in Canada is all over the place.

Rain storms, ice storms and snow storms are paralysing our transportation and electricity distribution systems. Both high and low temperature records are being broken everywhere. Cold snaps are persisting, snow is falling in places and in volumes seldom witnessed before; flooding is occurring widely. But, even with all this obvious evidence right before our very eyes, we still we find ourselves tip-toeing around the climate change issue. The public appears to still be confused or uncertain about what is happening.

Extreme Weather in Alberta & the Climate Change Question

So, what can we actually say we know?

The first thing we know is something we have known for thousands of years: human activities can result in climate change, even in the absence of greenhouse gas emissions. Civilizations that existed long before the widespread use of fossil fuels experienced the consequences of sweeping land-use changes. We also know that sweeping changes have been made to the landscapes of Alberta.

Since Alberta became a province in 1905, its landscape has been dramatically altered. Dr. Brad Stelfox has demonstrated that, in just over a century, southern Alberta in particular has been largely transformed. The region's hydrology has largely been altered through the construction of storage dams, engineering works and extensive irrigation systems. Careful control of water combined with a period of relative hydro-climatic stability has over time permitted ever more extensive and expensive developments to be built in southern Alberta's floodplains.

Alberta's vulnerability to extreme weather events has also been growing in step with its economic activities. Alberta is the highest greenhouse gas emitter in Canada and amongst the highest per capita in the world. So we see that Alberta has made itself vulnerable to flooding in two ways: the location of communities in floodplains, and through its contribution to changes in the composition of the global atmosphere that result in greater floodplain activity.

Think of Alberta as a pot of water on a stove. We are turning the heat up under the pot by constraining where water goes, through structural mitigation, to protect developments in floodplains. Then, we are putting a loose lid over the pot by way of climate change. Expect the pot to boil over.

Warmer temperatures are causing changes in the rate and manner in which water moves through the global hydrological cycle. Simple atmospheric physics decrees that a warming atmosphere can hold more water vapour which can serve as fuel for larger-scale rainfall events of longer duration.

The floods of 2013 certainly got everyone's attention. But what can we say we actually learned from them? What we have learned is that the math and the aftermath still don't add up.

When It Doesn't Add Up, We Need to Figure Out Why

What we are seeing is that our global hydrology is clearly changing. These changes are beginning to occur faster than our economies, markets and environment can adjust. The loss of hydrological stability is cascading through our climate system undermining the stability upon which our economic system depends for its stability. In the aftermath, however, we are beginning to glimpse how the math is changing.

Game Change 1: We need to begin controlling the hydrological cycle regionally and globally.

We know that hydrological conditions on this planet have always been changing. We have been fortunate to have had a century or so of relative hydro-climatic stability. That era, however, is over. The long-term hydrologic stability of the climate we experienced in the past will not return during the lifetime of anyone alive today.

This game change means that simply managing water in ways that are useful to us at a local scale will no longer be enough. We certainly have to be alert to changes locally, but we now also have to keep an eye on changes in the larger global hydrological cycle and (where possible) try to manage and adapt to them. This is a huge new concept – a societal game changer – and it is going to take time to get our heads around not just environmentally but economically.

GAME CHANGE 2: WHAT WE HAVE SEEN IS WHAT WE ARE GOING TO GET.

Predicted rises in temperatures of between 2°C and 6°C would result in further amplification of the hydrological cycle by 15% to 40% percent or more. This game change is not going to go away.

Because of our fur trade and colonial history, many Canadian towns and cities are located on flood plains in river valleys. According to a recent survey, some 20% of Canadians believe they live on or near what they describe as a flood plain; defending or evacuating these areas will be very expensive.

GAME CHANGE 3: SOONER OR LATER, WE HAVE TO ADDRESS THE ROOT CAUSES.

The new normal – as anyone who understands basic atmospheric physics knows – is that there isn't a new normal.

Unless we want our future to continue to be a moving target, sooner or later, we will have to confront the fact that we are rapidly altering the composition of our planet's atmosphere with significant effects on hydrology. This is a subject not many want to talk about in meaningful terms, even in the aftermath of the most damaging flooding in Canadian history.

GAME CHANGE 4: WE CAN'T DO THIS OURSELVES – WE NEED NATURE TO HELP US.

Structural engineering solutions are going to be necessary, but they are not going to be enough. We cannot ignore the local value of natural ecosystem processes.

In order to retain even partial rein over the hydrological cycle, we have to enlist all the help nature can provide us. We gain that help by protecting and restoring critical aquatic ecosystem function locally by and reversing land and soil degradation wherever we can. The importance of this cannot be underestimated.

Game Change 5: Managing water on a basin scale is no longer an option.

In this context, the watershed basin is the minimum unit at which water must managed. This should be perceived as good news because, at the basin level, the individual has the greatest power to affect positive change.

Conclusion: Don't panic – lead the way instead.

So, what now? The first thing we have to do is realize that the loss of hydrologic stationarity is a societal game changer. That said, this is not the time to panic and throw up our hands in helpless despair. The sky is not falling and the world is not coming to an end.

This is not a time for pessimism or resignation. It is a time for courageous and relentless citizenship and heroic leadership. We have to figure out the new math and make it add up in a changing world. In doing so, we will offer hope which in the future may become as precious to us as water.

About Bob

Bob Sandford is the EPCOR Chair of the Canadian Partnership Initiative in support of United Nations "Water for Life" Decade, which aims to inform the public on water issues and translate scientific research outcomes into language decision-makers can use to craft timely and meaningful public policy. He is also an associate of the Centre for Hydrology at the University of Saskatchewan and a Fellow of the Biogeoscience Institute at the University of Calgary.

[2]

Integration and Consciousness – Suzanne King

Boundaries don't protect rivers, people do. - Aristotle

MY JOURNEY FROM AUSTIN TO BRISBANE to pursue my master's degree in Integrated Water Management (MIWM) has been a tremendous opportunity for me to learn about the shift in consciousness that is occurring in terms of what we understand about water and its uses.

I moved from Texas, where our capital city is economically booming but has simultaneously experienced its worst drought on record, to the Australian state of Queensland, which recently also had its own worst droughts and floods. Coincidentally, Texas and Queensland became 'sister states' shortly after my arrival.

States in the U.S., like Texas and California, are looking to Australia for lessons to manage dwindling water supplies. Due to climate extremes and water vulnerabilities throughout Australia, the country has led the way in setting new precedents in the world for reforms, management and behaviour towards natural resources.

So what can we learn from Australia in terms of water ingenuity? Well, the solutions are often complex and intensive, but it's important to note that change is possible. Change takes immense amounts of resources, including human and financial capital to make high-level reforms happen. But real change occurs more effectively with community involvement and commitment, awareness, conscious leadership and meaningful communication.

The willingness of leaders to provide guidance and share best practices is not a new concept, but is further enhanced with collaborative communities. Rather than being closed due to fear, competitiveness and top-down approaches, transparency can foster true progress. With government transparency and the sharing of data and information, we can promote a purpose that enhances the well-being of the earth's water resources for the long term.

INTEGRATION IS THE KEY

People of different cultures and disciplines integrating holistic systems in harmony with our natural environment can lead to long-term water sustainability.

Integration can help us prevent costly mistakes and inefficient processes that historically may have been attributed to a lack of proper ecosystem and social impact valuations, operating in silos, a one-size-fits-all or megaprojects approach. Novel methods, such as new economic models, optimization and advanced valuation of ecosystem services, and integration of scalable systems, are being considered as potential solutions that may be better suited to meet the needs of urbanization in sustainable ways.

Although local needs and wide-scale impacts can vary, there may be principles that we can draw upon and integrate across specific contexts. For example, the Cooperative Research Centre for Water Sensitive Cities, formed in 2012, has a goal to "undertake interdisciplinary research that will revolutionize water management in Australia and overseas" (2014). The three pillars of a water sensitive city are: "access to a diversity of water sources underpinned by a diversity of centralised and decentralised infrastructure; provision of ecosystem services for the built and natural environment; socio-political capital for sustainability and water sensitive behaviours" (Wong & Brown 2009).

Water Sensitive Urban Design principles encompass similar attributes to a water sensitive city, such as storm water harvesting and permeable surfaces to lessen the environmental degradation caused by urbanisation, that have local water management to support environmental sustainability (Pilgrim 2007, p. 102). Water reuse is a key component to these overlapping, guiding principles.

Australian water reform, such as the National Water Initiative (NWI) 2004, was created "to achieve a nationally compatible market, regulatory and planning based system" to manage urban, rural and environmental water use and improve the triple bottom line (economic, social and environmental) (Commonwealth of Australia n.d.). As part of the NWI 2004, Water Smart Australia was one of the key initiatives formed with goals that included increasing "cost effective recycling and reuse of storm water and wastewater" (Pilgrim 2007, p. 79). The Australian Guidelines for Water Recycling (AGWR) were created in 2006 and augmented in 2008 to provide national frameworks for water reuse.

Some may argue that Australia's advanced treatment methods, as well as its stringent water quality requirements for non-potable reuse, is what has helped garner community support. For example, Coomera Waters located on the Gold Coast, provides the community development with recycled water for non-potable uses through dual reticulation, and has emulated natural ecosystems with methods such as bioretention systems, a constructed wetland and lake to filter stormwater pollutants and provide flood management benefits (see photo below).

Photo taken by Suzanne King.

The constructed wetland and lake pictured here, is located in Coomera Waters, which is "part of Gold Coast Water's Pimpama Coomera Waterfutures Master Plan and incorporates best practice management of stormwater runoff, wastewater systems and household water supply" (Healthy Waterways 2011). For more information, visit http://waterbydesign.com.au/coomera-waters/.

This growing awareness and gradual acceptance of integrated systems approaches and innovative policy reforms leads to further consciousness about our day-to-day interactions with water.

Raising Our Consciousness

> Together we create our future reality, so we should do so consciously, collaboratively, and responsibly.– Conscious Capitalism, by John Mackey & Rajendra Sisodia

Population and urbanisation are growing at unprecedented rates. While Austin is reportedly the fastest growing large city in America,

Australia is one of the most urbanised nations in the world, where 90% of its population lives in cities (Carlyle 2014; CSIRO 2014). Estimates state that 70% of the global population will live in cities by 2050 (CSIRO 2014), so this is all the more reason to be conscious of urban development as it relates to water.

With increased media coverage and water management concerns across the U.S., people are likely more aware of water issues than ever before. However, in places like Australia, severe water restrictions and extreme climate variability have already made people highly attuned to their water usage.

Continuing to raise awareness and working towards improving the social and environmental bottom lines can help avoid costly mistakes, such as consequential environmental mitigation and detrimental socioeconomic pressures. For example, according to a 2014 *Energy Policy* journal article, "even before accounting for negative impacts on human society and environment, the actual construction costs of large dams are too high to yield a positive return" (Ansar et al. 2014).

Our water policies, as well as our energy sources, food production, land use and personal consumption, can also have detrimental effects on our water resources. As consciousness of the energy, food and water nexus is increasing, it is evident that water management is no longer isolated. What happens in one part of the world effects another, and what happens in a distant part of the world can affect our local community. Therefore, rather than being separated by sectors, industries, neighbourhoods or political viewpoints, we can share a common purpose of protecting our waterways.

Water utility companies, such as Flow Systems, are proving how conscious business initiatives combined with effective governing mechanisms, such as the New South Wales Water Industry Competition Act (2006), can help deliver multiple benefits for public sectors, residents, developers and the environment in ways that meet the triple bottom line. An example is Central Park Water, where wastewater, rainwater and stormwater are able to provide the mixed-use urban development in Sydney with fit-for-purpose water (Flow Systems 2014). 'Fit-for-purpose' water usage, a WSUD principle, emphasises the need to consider the suitability of water (BMT 2009). In addition to potentially safeguarding ecosystems from polluted runoff and capturing valuable stormwater and wastewater resources, licensed operators can ease the demand on central systems and associated costs of large-scale infrastructure for urban development.

Awareness is continuing to rise from Austin to Australia. Systems can be viewed holistically and solutions can be implemented that emulate the natural environment as well as encourage local participation, long-term jobs and improved livelihoods. Fortunately, what is environmentally and socially sound is increasingly proving to be economically beneficial. For example, according to the *Bond Financing Distributed Water Systems* report from Ceres, Philadelphia's innovative water management project "found that a $1.2 billion investment in green infrastructure could achieve the same pollution control benefits as a $6 billion investment in traditional gray infrastructure" (Leurig & Brown 2014). Philadelphia's Green City, Clean Waters Program may also deliver employment, health, microclimate and property value benefits (Walshe 2013).

Sustainable methods for developing our cities, which are environmentally, socially, culturally and economically profitable are proving to be possible. Along with effective governance at each level and collaboration towards this shared goal, we are capable of advancing appropriate technologies and practices for today and the future. Some of these strategies include conservation, urban agriculture, green or living walls, stormwater and non-potable reuse, and resource recovery from wastewater, for multiple benefits.

Together, through integration of diverse systems and perspectives, and by raising our consciousness to create win-win scenarios, we can transform the story of our water future.

Acknowledgements

Many thanks to Chris Peacock for making this book possible, and to water heroes around the world who I've met, and who teach and inspire me from afar. Special thanks to the International WaterCentre, its founding universities, staff and professors for the opportunity to learn such complex topics in meaningful and pragmatic ways through the Master of Integrated Water Management and for such amazing classmates (from 28 different countries) who I continue to learn so much with and from. Thanks to the International WaterCentre Alumni Network (IWCAN) for working to support IWM practitioners globally.

About Suzanne

Suzanne King moved from Austin, Texas to Brisbane, Australia to pursue a Master of Integrated Water Management (MIWM) at The University of Queensland. She plans to utilize the interdisciplinary skills and knowledge gained through the MIWM program to advance sustainable water practices through meaningful communication and business development approaches.

Endnotes

Ansar, A, Flyvbjerg, B, Budzier, A, and Lunn, D 2014, 'Should we Build more large dams? The actual costs of hydropower megaproject development', *Energy Policy*, vol. 69, pp. 43-56.

Commonwealth of Australia n.d., *National Water Initiative*, Australian Government Department of the Environment, viewed 1 September 2014, http://www.environment.gov.au/topics/water/australian-government-water-leadership/national-water-initiative.

BMT WBM Pty Ltd (BMT) 2009, *Evaluating options for water sensitive urban design – a national guide*, Joint Steering Committee for Water Sensitive Cities (JSCWSC), viewed 5 June 2014, http://www.environment.gov.au/system/files/resources/1873905a-f5b7-4e3c-8f45-0259a32a94b1/files/wsud-guidelines.pdf.

Carlyle, Erin 2014, 'America's 20 fastest growing cities', *Forbes*, 14 February, viewed 28 August 2014, http://www.forbes.com/sites/erincarlyle/2014/02/14/americas-20-fastest-growing-cities/.

Commonwealth Scientific and Industrial Research Organisation (CSIRO) Water for a Healthy Country Flagship 2014, *Urban water: transforming urban water systems to deliver sustainable, productive and livable cities*, CSIRO, Australia, viewed 4 September 2014, http://www.csiro.au/urbanwater.

Cooperative Research Centre for Water Sensitive Cities (CRCWSC) 2014, *About the CRC*, viewed 4 September 2014, http://watersensitivecities.org.au/about-the-crc/.

Flow Systems 2014, *Central Park Water a local sustainable community*, viewed 5 September 2014, http://centralparkwater.com.au.

Healthy Waterways 2011, *Water by design Coomera Waters*, viewed 4 September 2014, http://waterbydesign.com.au/coomera-waters/.

Leurig, S & Brown, J 2014, *Bond financing distributed water systems: how to make better use of our most liquid market for financing water infrastructure*, Ceres, Boston, viewed 7 September 2014, http://www.ceres.org/resources/reports/bond-financing-distributed-water-systems-how-to-make-better-use-of-our-most-liquid-market-for-financing-water-infrastructure/view.

New South Wales Government (NSW) 2014, *The Water Industry Competition Act* (2006), no. 104, viewed 4 June 2014, http://www.legislation.nsw.gov.au/maintop/view/inforce/act+104+2006+cd+0+N.

Pilgrim, JJJ 2007, *Australia's water resources: from use to management*, revised edn, CSIRO Publishing, Collingwood, Victoria.

Walshe, S 2013, 'Philadelphia water management: from grey to green infrastructure', *The Guardian*, 11 June, viewed 19 September 2014, http://www.theguardian.com/sustainable-business/philadelphia-water-management-green-infrastructure.

Wong, THF & Brown, RR 2009, 'The water sensitive city: principles for practice', *Water Science & Technology*, vol. 60, issue 3, pp. 673-682.

[3]

The End of Traditional Conferences: How water discussions must evolve – Karen Kun

WATER IS A COMPLEX TOPIC... but we still need to talk about it.

We've seen how water authorities or organisations start with the same exact conditions, but they seem to produce different outcomes – this is because of the variety of interactions within the elements of the system. And this will happen more often, unless we start new ways to discuss water issues and establish some stability in our systems. By doing so, we can lessen the complexity of water itself, and we can begin to expect - even rely on - several outcomes in the future.

With my long-time friend Tatiana Glad, I was looking for an opportunity to contribute our skills in facilitation to the water sphere. We identified a major gap in the outcomes from water events we attended, and we noticed that more collaboration and more innovation happened when water experts were truly involved in a balanced discussion. It's also important that these discussions allowed everyone to include different perspectives and individual pieces of wisdom as it related to the overall big picture.

This premise became the foundation for Waterlution, the organization we created to offer a combination of water content <u>and</u> a facilitation process about water. Waterlution believes that only with the presence of both content and process could meaningful dialogue and strategic actions on water occur.

What is the significance of dialogue?

Traditional formats of professional gatherings, such as panels, talks and lectures at conferences, have exhausted the value they can offer. These uni-lateral (or, at best, bi-lateral) forms of communication – we call them "knowledge download" sessions – rob participants from any opportunity to be active in knowledge production and exchange.

These gatherings create a dichotomy between "experts" and "non-experts"; at Waterlution, we believe everyone has some level of expertise to offer. We think that magic happens when new and unforeseen ideas are added to the mix on a topic because they create space for "A-ha!" moments to emerge. When more of these moments occur, more opportunities for knowledge co-creation and co-production will emerge in order to push the water innovation envelope.

How does Waterlution deliver unique dialogue experiences?

Waterlution's purpose is to inspire pattern-making (and pattern-breaking) changes in order to have a healthy, sustainable relationship with water. This is achieved through designing and delivering gatherings that take people on a guided journey to find their authentic water voice, which then builds their confidence to share what they know across sectors and viewpoints.

Starting from small gatherings around the dinner table, Waterlution has now transformed into an organization that involves people from a variety of sectors, perspectives and generations. We've seen how Waterlution creates a "cross-pollination" of ideas: the way we design, deliver and inspire (through facilitated dialogue) provides everyone the opportunity to dive into the complexity of a topic.

Participants start to understand that systemic change happens as individuals respectfully share knowledge – from professional experiences as well as personal or cultural knowledge – and as they develop broader perspectives through open dialogue and community building. Waterlution's core value of cross-sector dialogue allows for respectful conversation to take place around large-scale topics – from oilsands to fracking and agriculture – while providing space for all perspectives.

> This is an important component of building opportunities because we must have open-minded conversations to explore new ideas, perspectives and realities. Participants interact on an inter-personal level, instead of a formal/hierarchical level.

How can this help our discussions evolve?

The challenge with similar gatherings is keeping them interesting – through practice, we've embraced this challenge to maintain a steady flow of bold ideas without compromising the discussion to end up with the lowest common denominator. Waterlution is not a negotiation organization; we are an idea generation organization that's focused on water. Different viewpoints can address emerging concerns, spark thoughts of innovation, and help others find new ways to safeguard water resources.

In our sessions, we design everything from when people arrive into the "experience" until they exit, even though we expect (and encourage) the unexpected. We have the confidence and skills to stick with the group through the uncomfortable moments that emerge because those times of conflict and disagreement often turn into moments that hold the greatest value. Where many conferences and organisations try to avoid conflict, we carefully harness and use that energy to create positive discussions and outcomes.

Through these experiences, we encourage participants to "create the amazing" because we have the ability, resources, skills and education to do it; we also believe that it's a goal that people are attracted to. Recently, at our second Water Innovation Lab (WIL2013), Waterlution brought together young leaders and professionals for 6 days to disconnect from typical behaviour and have honest discussions about creating a dynamic water future.

5 THINGS WE LEARNED TO HELP WATER DISCUSSIONS EVOLVE INTO SOMETHING BETTER:

In autumn 2013, WIL2013 took place in the Canadian Rockies to get away from everyday comforts and broadband Internet usage. This was specifically to promote one of our design features of "disconnect from what you do every day and connect with who (and what) is present".

Our top five lessons from WIL 2013 were as follows:

1. MIXING

Similar to combining unknown ingredients to create something to eat, the combination of experienced guests and new participants will spark a lot of conversations and debates.

Bringing in participants and resource guests – at Waterlution, this is what we call our experienced guests – with not-so-obvious water connections took a lot of time, energy and personal attention... but it was worth the effort! Innovation comes from people asking interesting questions while being part of a unique shared experience. By mixing those people with creative discussions/debates, everyone leaves with ideas, companies and projects never before imagined.

2. BALANCE

At WIL 2013, magic happened in the spaces between chaos and order. The Lab design sought balance between: differing perspectives; the expected and surprises; comfortable routine and new challenges; as well as structure/schedule and play/openness for new ideas and connections to emerge. Opportunities for playfulness and time in nature were key components in the overall design.

3. OPPORTUNITIES FOR PRACTICE

Learning to practice, and practising what we learn, became a crucial part of WIL2013.

Participants of "Stream 2 – Strategic Conversations in Water Work" played a pivotal role in closing the Lab. By providing opportunities for practice, we witnessed their enjoyment, experienced their skills and celebrated in their confidence.

4. BUILDING TRUST:

Trust in one another and trust in the process.

From our experience, we know it's important to spend several days and nights together to build trust among participants, most of whom were strangers prior to the Lab. There is something exhilarating about being surprised.

Before the Lab, everyone was pushing for an agenda, wanting to know what was going to happen when and where. We provided enough of an agenda for participants to come; yet we purposefully left a number of unknowns. Quickly, participants trusted us enough to convene despite not having all the answers of what the Lab entailed.

5. CHALLENGING:

What's an experience without a challenge?

Often, the Lab was very much about challenging participants to step outside of their comfort zones. As hosts and conveners, we also challenged ourselves to do something we had never done before in the Lab design – if we were asking others to embrace layers of discomfort for transformational learning, then we needed to embrace the same ideology.

So what's the end result?

Participants are now thinking from a new perspective, and they are considering ideas they had never even thought about earlier. People commend us for designing respectful spaces where vastly different viewpoints can co-exist in one room; when we don't vilify others' opinions, we can actually learn from one another and begin to collaborate on projects.

The best part, for us, is to watch numerous young leaders remain focused on water, as they develop their career leadership presence in the water space and integrate more Waterlution-style practices into what they do as they become "waterpreneurs".

About Waterlution

Waterlution recently celebrated its 10-year anniversary, and the organisation has accomplished quite a lot through delivering over 60 workshops held in nearly every Canadian province and territory. Over its 10-year period, Waterlution has had the participation of 3,500 young leaders, the training of more than 50 Associates, and local hub networks are more involved when communities need to talk about water concerns in major cities.

Karen Kun and Tatiana Glad (co-founders of Waterlution) had met in Montreal during University as active members of AIESEC, both studying for their commerce degrees. In addition, they had a shared passion for the natural environment, and they saw how business could be a positive driver toward sustainability. After a chaordic design process in Zurich, Karen became Executive Director of Waterlution while Tatiana moved to Amsterdam, and became Karen's more trusted listener/supporter.

Karen would spend the better part of the next few years travelling across the country, meeting with experts, and learning at events and building a network with practitioners living deeply with water challenges across a host of sectors.

[4]

Developing Leadership in the Water Sector – Renee Kayal

OUR FIELD IS CHANGING ALL THE TIME, which is why we need to ensure that we have the leadership to push us through the challenges in the future. In a sector dominated by detailed, focused professionals – often with hard science backgrounds – there is a need to identify the skills and qualities that will develop leaders within the water industry.

I've looked though my personal interactions, stories of others and case studies to evaluate the challenges that are reported every day. This list might seem a bit off, but I see these skills acknowledged and encouraged as key components in the professionals that will serve the ever-changing water sector:

COLLABORATIVE

Be willing to work with those above, below, and next to you. See what you can learn from someone on the way to your goal and envelop them in the process. Provide kudos, honor team effort and be willing to work *with* a team, not just lead a team.

Flexible/Adaptable

Change is inevitable; learn to embrace change or at least expect it. Prepare yourself to pause in the face of modification and relax, rather than charging ahead. Be remembered for how you acted both in the face of adversity and in the good times.

Communicative

Connection and communication are key to all interactions. Acquire the ability to craft your message to diverse groups and be confident to communicate in various forms. Communication is two way – be sure you master the ability to listen.

Innovative/Creative

Do what you do in new ways. Open yourself to wandering down new paths, and review repetition. Ingenuity bubbles up all around us; learn to foster and explore rather than suppress.

Determined/Dedicated

Passion is important. Believe in a vision and your piece of that vision. Work to light passion in others rather than pulling or pushing them to your vision.

Connected

Learn from others and grow your sphere of connections. Experiences and opinions of others are invaluable to you. Look to the interrelatedness of nature and embrace a holistic approach to learning and delivering.

THE WEF LEADERSHIP CHALLENGE

In early 2011, a group of us at the Water Environment Federation (WEF) were challenged to develop a leadership program that would attract professionals from across the water sector. We veered from the original idea because we began to focus on the importance of connecting peers so they could discuss leadership issues and skills. We thought that focusing solely on skill acquisition seemed like a short-sighted solution for a long-term concern.

This thought process was just the first step into the creation of the Water Leadership Institute. By attracting participants from all areas of our sector from all professions and ages, we could guide them through a structured program (between seven to nine months) that would open them to new ideas and experiences, while building long-term relationships with their colleagues.

"The Water Leadership Institute program is aimed at educating, training, and providing opportunities that enable developing and emerging leaders to build strong lasting relationships within the water industry."

That was our mission statement, so we worked hard to roll out a program to accomplish those tasks. We focused on key fundamentals, such as emerging trends, public relations, entrepreneurship and innovation. Then, to add realistic complexity to this program, a final team project would examine the challenges facing water scarcity/delivery from different stakeholder perspectives. Experts would hold online discussions about specific topics, host group conference calls for project development and interact in an all-day workshop at the end of the program.

After launching the program, we were absolutely thrilled to receive a strong response of interest, shown by the number of applications we received from individuals across the United States and in several other countries. We selected a class that included a diverse range of professionals from utilities, academic institutions, consulting firms and nonprofits. In addition to their respective credentials, many candidates had already experienced some type of leadership training. This commonality first clued me in to one component of our vision for the Water Leadership Institute (WLI) where people want to establish strong, long-lasting relationships within our sector.

Our inaugural class completed four webcasts, a sparsely-attended face-to-face event, and the all-day workshop. The group remained enthusiastic throughout the event, and the culmination activity was a high point for many; even with a successful session, we sought extensive feedback on how to improve the 2013 WLI.

In addition to expected topic and logistics suggestions, most feedback focused on improving participant interactions with speakers through more face-to-face time – both online and in-person – as well as more opportunities to interact with each other. Most attendees listed in-person collaboration and networking as the most beneficial part of the program, where they felt invigorated by the last day of the event because it allowed them to move from virtual networking into real-life networking. The new graduates called for continued forums for interaction and participation in the water leadership program, which included plans for an alumni network, the ability to serve on the program steering committee and invitations to attend future workshops and graduations.

Using Leadership Qualities to Improve Leadership Training

As we were creating the program syllabus and overall plans, we were nervous but misdirected our energy. We were too focused on the content, which led to establishing relationships as a secondary goal – silly us! Content was simply the vehicle to help participants establish the connections that our participants craved.

By using my brief list of leadership qualities, we needed to reflect on the feedback and find creative/innovative ways to adapt our plans. We may be considered the "leaders" of the WLI, yet this doesn't mean we can't adapt, collaborate and learn just like our participants.

So we listened and worked on new options to add flexibility (and a controlled amount of uncertainty), such as turning the one-way webcast component into a two-way interaction between students and speakers, forming the teams for the final project much earlier in the program, and adding an optional day-and-a-half meeting at the WEF. The in-person meeting would focus on communication skills, input from local leaders, and we would hold a discussion on new ideas around green infrastructure.

In addition to the new elements, we were determined to incorporate a collaborative session between students and alumni, which would enhance networking opportunities. We planned a graduation celebration at the end of the culmination workshop, where the WLI 2012 alumni, speakers and guests would be invited to communicate in person with members of the WLI 2013 class.

From Leadership Student to Leadership Trailblazer

With ten WLI graduates forming a new program steering committee, they helped make the 2013 program an even greater success, especially because they could evaluate the projected plan from a student's perspective. The WLI 2013 program had a clearer workflow plan for participants, it established the roles of the steering committee, and included a wider range of topics for discussion.

We held an interactive panel on leadership experiences, talked about utilities of the future, discussed stormwater/green infrastructure, and looked at entrepreneurship in the water sector. As discussed in our planning, the final project teams were set earlier, and we assigned each team a mentor from the 2012 class.

The full-day workshop was well attended, even with alumni and others exposed to the program who stopped by the event to observe. The camaraderie with team members became very clear, presentations were completed by each group and discussions were freer flowing; I attribute this to the fact that the groups knew each other well and felt comfortable expressing themselves.

Improving Leadership Training in the Water Sector

The WLI 2013 class stressed that they only really got to know the few folks in their final project team and wanted even more opportunities for connections. They also called for more "soft" topics and experiences, such as what industry leaders look for in team leaders and hosting leaders from outside the sector. Other ideas revolve around alumni sessions and networking opportunities, including discussions during the final workshop.

These are the continuous steps leaders and trainers need to take in order to keep our standards high. A bland, generic and monotonous leadership program has a lot of modern-day limitations, not to mention how it will likely produce bland and generic leaders.

> By taking a proactive approach to communicating and collaborating with leadership students, we can learn to improve our training techniques and help make those innovative changes in our future leaders.

With additional improvements like video-based work groups and alumni experience sessions, we held WLI 2014 in April with a very diverse class of 27 participants. Our Steering Committee was updated, as new alumni wanted to participate, and it is currently composed of three WLI 2012 alumni and seven WLI 2013 alumni. The application process for WLI 2014 was marketed with a more aggressive strategy, and we were rewarded with outstanding applicants. The review process was detailed and difficult, but we have learned quite a lot over the past three years on improving the leadership sessions and our preparation to make them successful.

That brief list of leadership qualities – from the start of this piece – was a result of my observations during these past WLI sessions. Of course, I have seen evolving examples in my years in the water sector, but the intimate view I get into the paradigm of each year's WLI participants is clarifying and inspirational. These individuals are motivated, and they share a view in improving and innovating the water sector, as if they are all facing the same massive challenge as one team.

These participants are just a microcosm of what the water sector should expect in the future of its leadership. Our leaders want to have relationships with others (who have differing opinions) to build on their personal and professional views. They are looking at water and challenges in a very broad sense and find excitement in knowing someone on a different edge of the same plane. I dare each of you to look around, not merely focused ahead, as we foster on a journey to a smarter water sector – perhaps the Water Leadership Institute will be a big step in your professional journey, too.

About Renee

Renee Kayal serves as the Education and Training Program Director for the Water Environment Federation. She has been with the WEF for over ten years in various roles, primarily in the technical and educational services group.

[5]

When Women's Networking Gets Personal (in the good way) – Leila Munla

IN THE RIGHT CONTEXT, your professional network can become your personal network, and that could potentially change the future of collaboration and dialogue for women in this sector.

Traditional networking includes activities that would be considered relevant or directly connected to a particular field; any fun or challenging activities that aren't related to work should be kept private. However, many of these "inappropriate" or "irrelevant" activities take place at corporate bonding retreats, which aim to improve working relationships and establish stronger bonds with your colleagues – so what's really the difference?

Whether you consider it networking or corporate bonding, any activities outside the scope of our field gives us the opportunity to share a part of our personal lives with some of our colleagues. We all have hobbies and interests, plus we all have hidden talents that could be improved by sharing them with others. This is why establishing a professional network with a personal mindset is beneficial for your career as well as your life outside the office.

How I Found This Hidden Gem of Women's Networks in the Water Industry

For me, it all began on a chilly November afternoon in 2012 while I was having a conversation at the Water Quality Technology Conference in Toronto. I was talking to Jane Simmons about the rewarding experiences of volunteering with women, while mentioning how there seemed to be a lack of local networking opportunities for women in the water industry.

After bouncing around some ideas, we decided to gauge local interest by hosting a small potluck dinner with a good mix of eight professionals and graduate students.

> *This group – we lovingly called WOW (Women in Ontario in Water) – created some fundamental concepts and determined that it's time to move past networking events that were simply strangers in a room full of more strangers.*

This led to the idea that any new member would need to be recruited by an existing member, which would keep WOW as a small, close-knit group and would allow recruits to be properly welcomed by other members. The concept wasn't to create an exclusive club with arrogant rules, but we wanted to cultivate strong relationships by keeping the group numbers low instead of expanding exponentially; this became a key factor that naturally transformed professional-only relationships into personal ones.

Monthly gatherings were another key factor in building personal connections within a professional organization. These "meetings" would alternate between potluck dinners and fun activities, which included soap making, rock climbing and canoeing. Since these women were actively volunteering to host potlucks and organize events, the group developed rapidly and we became very good friends in a short amount of time. It was quite difficult to walk away from this group of amazing women after completing my PhD.

However, it seemed like perfect timing after I received my doctorate and moved to San Diego. I met Dawn White a few months later at the Water Quality Technology Conference, and she was interested in my proposition to start a women's group in the water industry.

After several discussions, we decided that we would put together a workshop at the upcoming California-Nevada section of AWWA conference. Our "Women in Water Working Together" workshop was a chance for women from different backgrounds to share their experiences and begin a solution-oriented dialogue on improving the current state of the industry for women. We wanted to lay the foundations of a supportive platform, foster a sense of inclusiveness for both seasoned and young professionals, and our efforts would inspire other women to form local networking groups. Thanks to the tremendous feedback we received, this workshop will be a repeat event at the biannual conference.

Finding Support Beyond Typical Networking

Partnerships are built between real people, so the stronger the connection, the better the relationship. Also learning new skills together, outside of the typical networking activities, can help you both grow on a professional and/or personal level. Women may find this approach to be more natural as we tend to intrinsically share tips and tricks, we're comfortable being empathetic and supportive, and we also tend to share our experiences. Industry-specific conversations will happen naturally and information will undoubtedly be shared, but they do not need to be the focal point.

With gender-specific issues in nearly every industry – proving yourself competent, differences in pay, and having to choose between family and career – it is necessary to have female-oriented groups to provide a supportive network. These groups are particularly important in industries that are still dominated by men, such as engineering or computer science.

Sure, we go undercover and try to join the "brotherhood", hoping we won't be discovered for the women we are. We might have to fight harder and prove ourselves, just to earn respect that may normally be granted right away; we'll likely hear "no" more than "yes".

Yet, I think in our strife to become and define ourselves as professional women, we have tended to desperately excise the very part of us that makes us women. We have followed the rule of "if you can't beat them, then join them", but we are not very good at being something we are not. We will always have these challenges, and they are not simply going to go away if we continue to deny our feminine birthright. And, yes, I say birthright because we cannot forget that our unique perspective is a genuine contribution to the industry.

The Change in Women's Networking Starts from Within

Instead of feeling like an outsider and struggling to be heard, what if that water professional inside you was automatically inaugurated into a group of women that would support you, mentor you and nurture you along your career journey? Then start with yourself – and throw your traditional concept of a professional network out the window!

Wouldn't it be exciting to get together with like-minded women and learn about new hobbies *while* you're networking and building new partnerships? We do not – and should not – have to be identified by our career or our professional affiliations, so neither should be one dimensional.

It is true that there are already several women-only groups, such my first one called "Women in Engineering". The experience was always gratifying and offered a wonderful platform to foster a sense of support for professional women, but I would like to argue that they are not enough; they need to be industry specific. The rewards of a non-traditional format within an industry-specific women's group could help transform our industry.

The women who are part of these supportive networks are not only ambassadors for the industry today, but are on the path to be the leaders of tomorrow. Can you imagine where all these women will be in 10 years? Not only will they help guide our water industry toward a better future, but they will proudly reflect on the support they received to get to this point in their careers.

This is beneficial for the entire industry because strong women, who have a passionate and active interest in the topic, will work hard to improve the future of *all* parties involved. The future of water will have conflicts, uncertainty and economic instability, but a foundation of strong partnerships between women within the industry can help ease those tensions.

I can envision how the relationships forged in these networking groups will breed a collaborative spirit, which will create an environment where women can perform their best and contribute the most to the water industry. By making sure there's a strong foundation – between men or women, small businesses or large producers – the future of water will have less arguments and more solution-oriented discussions.

And we all benefit from that.

About Leila

Leila Munla recently completed her PhD in Environmental Engineering at the University of Waterloo in Ontario, Canada. Her research focused on the fouling of ceramic ultrafiltration membranes in drinking water treatment. She now lives in San Diego and continues to passionately work with women in the water industry.

[6]

Water Issues are People Issues – Julia Ko

"You environmentalists throw a lot of misinformation out there. Our technologies protect water."

THIS WAS SAID TO ME BY AN EXECUTIVE who was referring to a method of extracting oil, using steam instead of conventional mining techniques. I just finished a presentation on successes of businesses and environmental groups that work together on water issues; the room was silent except for one person who wanted to grill me about water and oil.

This presentation – at a long mahogany table of men at an exclusive club, some of whom are energy executives – was intended to open a cooperative conversation. In theory, his point is valid that the technology does use less water, but it doesn't necessarily "protect" water. I reacted hastily to his badgering, and I shouldn't have.

"I am sorry if this presentation was inappropriate for this audience. My job involves difficult discussions, and if you wanted a feel good presentations, then I would have shown you pictures of water wells and children in Africa."

Not my proudest moment. My bad.

THE DEATH OF ENVIRONMENTALISM

My audience had a heightened level of sensitivity to all groups who claim to be focused on environmental issues, especially after a hot media campaign that dealt with conflicts between both sides. So, with hesitation and uncertain feelings about these groups, that would understandably make the discussion of "cooperation" unpalatable.

This experience let me live the *Death of Environmentalism*. I realized that being an "environmental group" was a toxic situation for conversations about the trade-offs between energy, water and food. Yet, when I became a "conservation group" looking for more options to increase water use for economic and environmental purposes, I wasn't immediately discarded or considered a tainted enemy.

It's odd to think that certain terms or phrases are now threatening to specific individuals, even though the focus is helping others; I guess it depends on their respective agendas. But that conflict of mindsets is a microcosm of how water issues – in addition to other environmental concerns – are basically people issues.

It doesn't matter if a particular technology uses less water when the end result is poisoned water. Some people don't want to discuss water issues because they expect someone else to handle the problems... until it's too late because no one was willing to have an open conversation about water.

WATER IS LIFE & ITS ISSUES ARE OUR ISSUES

My work at Water Matters gave me a first-hand look how water involves more than just planning or logistical issues, such as energy, drought planning or flood preparations.

I consider each issue like a park user. There are dog owners, parents, toddlers, school kids, university students and so on. Each user enjoys the park for different reasons, whether you're taking a jog with the dog, you're a parent who wants the kids playing outside instead of on the sofa, or you're a high school kid who is hanging out with friends.

Water is used for different reasons with different intentions. It's still the same water – or, in this metaphor, the same park.

The park is the background and how we use the park is what matters. For example, there is a park in Vancouver that is currently gathering public input on a long-term plan, and dog owners seem dissatisfied with a fenced area that's relatively small. I am sure many readers are familiar with these types of disagreements, where one user unhappy with a plan that other users say they want, but no one is willing to talk to each other. This turns into a really complex discussion with a variety of unspoken viewpoints.

Do you see the comparison to water?

We're no longer talking about people who use the park... it's just about people. Water issues are much more than the individual uses of the outspoken water users, so we need to make sure that there's a steady dialogue about water from all types of users. Similar to the park metaphor, the loudest protestors aren't always resolving the issues for the silent ones.

From the Life of Water to the Life as a Parent

When I left Water Matters, I thought I understood the complex nature of water policy.

HA! Wrong again... giving birth to my first son changed my life significantly in that aspect, even if it's another human thing that doesn't seem remotely connected to water-related issues like aggregate extraction and land-use changing groundwater connectivity.

For my son, I moved from Alberta to British Columbia, leaving my value as a water policy person in Alberta to a province that doesn't necessarily value my skills or need my expertise. My son made me reflect on my work and *how* I worked; in the past, 60 hours a week was typical behaviour.

I wanted to be a parent first, then fit in time for my career. After some introspection, I came to the conclusion that if I couldn't work on water issues because of competing demands for my time, then maybe other people couldn't give as much time to the same issues that I cared about.

This is why we need to focus on water issues with the mentality that these are truly human issues, and they won't go away because we're not paying attention to them.

> *From old infrastructure to water conservation, these issues will affect people more than ever – especially our children and grandchildren – so we need to have those awkward conversations if we're going to make a real impact.*

ABOUT JULIA

Julia is the mother to Oliver Ko Rock, a two-year-old toddler, and she is a water warrior who follows conversations in this sector that are of interest and importance.

[7]

Advancing One Water Management with One Water Communication – Donna Vincent Roa, PhD, ABC

WATER IS A MULTI-FACETED GOOD and a global priority that affects human beings, as well as the survival of institutions that rely on it. Those interested in shaping the future of water management push the notion that **"water is water"** to encourage comprehensive management of our waters on a transformational scale and push the industry towards a "one water" resource management. However, to make water everyone's business *while* achieving this goal, we must also have **one water communication**.

Engineers, scientists and communication professionals who serve the water industry must address the communication challenges with the topic of water. A holistic approach will shape the thinking of the current generation, stimulate ownership and enhance awareness about the sustainable use (and economic importance) of water. We need to teach people how to value our water resources by regularly providing positive storylines about water, while including the cycle of water in concepts, images and a language that they can understand.

Charles Fishman, author of *The Big Thirst*, mentions that "water expressions infuse our language, but we don't really have much of a language or a framework for talking about water itself." He says that our success with water has allowed us to become water illiterate, and that "our everyday attitude about water is filled with contradictions."

> "We are on the verge of a second modern water revolution – and it is likely to change our attitudes at least as much as the one a hundred years ago. The new water scarcity will reshape how we live, how we work, how we relax. It will reshape how we value water, and how we understand it," Fishman wrote.

Language directs and influences our thoughts as it helps us to understand concepts of time and space. Language affects how we conceptualize the world and plays an important role in shaping our realities, perceptions and understanding. Language that's coupled with standardized visual cues – like colors and illustrative icons – can create sensory representation to support linguistic representations. By using all three – language, icons and colors – we can shift our understanding of water and its value to society.

With this combination, we can influence cognitive processes (e.g., thought and experience). While I am no expert in the principle of linguistic relativity, I do know that we are often at the mercy of our language and symbols. This puts pressure on those who communicate about water to take extra care in the language, signs, symbols and colors we use.

One water communication should include more than just developing a common language – we need a visual conceptual framework. Our industry needs a framework that creates an exact and more consistent communication about water.

We need to create awareness about water issues because informed perceptions increase public confidence; that's good for utilities and the public. We need to use communication to develop a universal water ethic and change behavior at the citizen level towards our most precious resource on earth. Standardized visual communication will aid us in the effort.

I am fully convinced that, for water communication, we need pervasive terms, labels, illustrative icons, colors and a simplistic categorization that correctly reflects the realities of water. Technical terminology and language alone do not facilitate the level of understanding that we seek. And, when it doesn't, it's easy for the media to "sloganize" the meaning of an idea (e.g., toilet-to-tap or the "yuck factor") with terms that disparage our sustainable water processes or erode the importance of our water stories.

We need to reclaim water from media-driven clichés and confusing buzzwords as we eliminate stigmatizing language. Additionally, it's hard to clean up or rename processes that have already been labeled with negative terminology or language with inherently negative connotations. I hesitate to include those terms because of the damage it does to our effort to educate the public on water issues and our ability to tell the full water story.

We must look beyond language we use to communicate about water. There is an overuse of engineering terminologies, technical buzzwords and stigmatizing language. Then there's a lack of a standardized color palette, and there aren't standardized or common icons used throughout the industry.

There's even inconsistent use of colors and icons representing the water cycle, water processes, equipment and so on. For example, while Irvine purple (Pantone 512; RGB 131,49,119; CMYK: 53,99,3,18; Hex #833177) was chosen as a color to differentiate between pipes that carry recycled water and those that carry drinking water, the variety of the colors in use by the industry range from pinkish to dark purple.

It's my belief that we value what we understand. Throughout history, we've used scales, models and frameworks to help us understand complex matters or materials. Consider the contributions of Daniel Defoe, Charles Darwin, Anders Celsius and John Smith Elgin Marbles when thinking about standardized methods to understand specific details.

My favorite is Rear Admiral Francis Beaufort, author and developer of the Beaufort Wind Scale. Study the naming of winds and you'll find that getting to a standard was not easy.

As hydrographer to the British Admiralty and an Irish Royal Navy officer, he developed the Beaufort Wind Scale, even though there were many wind scales in use at the time. The various scales led to much confusion, such as inaccurate measurements and poor communication about wind.

Beaufort succeeded in standardizing the way that someone could provide objective and concise descriptions about weather observations. The Beaufort Scale is one of simplicity and clarity with only four categories: Beaufort number (0-12), name, wind speed and description.

Midshipmen in the Navy learn the Beaufort scale, and sailing courses include instruction on it. The modern version of the Beaufort scale uses colors (from a light teal to orange and red), concise labels and descriptions, sea conditions, numbers and photos to indicate the level or force of the wind. Some versions include a variety of pictures illustrating the wind's force.

Beaufort's intentions were to shape the views of the world on wind. I believe that we have a similar opportunity to shape the views of the world on water.

I believe that we need to acknowledge the impact of language and make a concerted effort to provide a linguistic and visually accurate conceptual window into the world of water. The Roa Conceptual

Model for Water Communication™, its water icons and the color palette, provide a framework for standardized visual communication.

The model below visually represents the water treatment cycle in a simplistic, user-friendly way. Tested and reviewed by industry colleagues, water engineers and school children under the age of 16, this model clearly illustrates how water is water, how water is treated to different classes (or levels) of quality, when the cycle is closed (or looped), and how we need to understand that there are many resources to be extracted from water that has been used.

Throughout history, we've had an affinity for categorizations and using color to help categorize. All societies have social codes and systems based on color; from the Roman period through the Middle Ages, the church started to use color for symbolic, practical and categorization purposes, which continues in present-day design and communication. Color categorization is a familial concept and a universally understood framework that has the potential, when used correctly and consistently, to enhance understanding and carry messages.

My goal, with The Roa Conceptual Model for Water Communication project, was to develop a simple taxonomy and clear framework that would help to standardize visual communication for water. This could create the standard for water associations, water utilities, engineers, scientists and water leaders worldwide to communicate visually about water.

54 | *Damned If We Don't!*

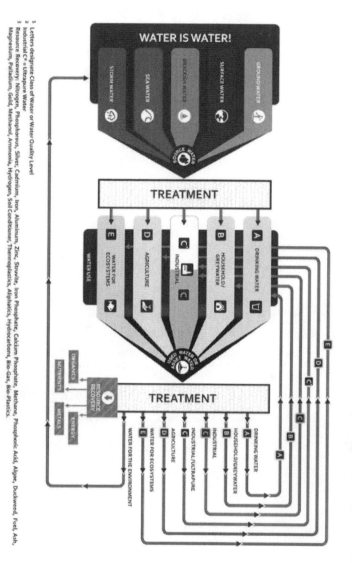

Ideas for Accelerating Change Around Water | 55

The Roa Conceptual Model for Water Communication™

Water Icons and Color Palette

TREATMENT
TEXTURE

HOUSEHOLD / GREYWATER
CMYK: 0, 0, 0, 15
RGB: 220, 221, 223
HEX: DCDDDF

WATER FOR ECOSYSTEMS
CMYK: 1, 18, 0, 0
RGB: 252, 213, 182
HEX: FCD5B6

GROUND WATER
CMYK: 66, 37, 0, 0
RGB: 89, 142, 202
HEX: 598ECA

BRACKISH WATER
CMYK: 27, 45, 100, 6
RGB: 184, 134, 11
HEX: B8860B

USED WATER OR ASSET WATER
CMYK: 0, 0, 0, 100
RGB: 35, 31, 32
HEX: 231F20

INDUSTRIAL / ULTRAPURE
CMYK: 0, 0, 0, 0
RGB: 255, 255, 255
HEX: FFFFFF

INDUSTRIAL
CMYK: 0, 0, 0, 35
RGB: 177, 180, 182
HEX: A6A6A6

SOURCE WATER
CMYK: 99, 82, 37, 27
RGB: 24, 56, 95
HEX: 18385F

RAIN WATER
CMYK: 5, 82, 0, 0
RGB: 250, 72, 184
HEX: FA48B8

SEA WATER
CMYK: 91, 49, 44, 18
RGB: 3, 99, 114
HEX: 036372

RESOURCE RECOVERY
CMYK: 85, 20, 100, 8
RGB: 29, 139, 67
HEX: 1D8B43

DRINKING WATER
CMYK: 20, 8, 0, 0
RGB: 199, 217, 240
HEX: C7D9F0

AGRICULTURE
CMYK: 16, 2, 31, 0
RGB: 215, 228, 189
HEX: D7E4BD

RECLAIMED WATER
CMYK: 53, 99, 3, 18
RGB: 131, 49, 119
HEX: 833177

SURFACE WATER
CMYK: 44, 72, 85, 58
RGB: 82, 47, 26
HEX: 522F1A

STORM WATER
CMYK: 53, 51, 100, 36
RGB: 97, 87, 33
HEX: 615721

[1] Letters designate Class of Water or Water Quality Level
[2] Industrial C* = Ultrapure Water

© Donna Vincent Roa | www.DonnaVincentRoa.com | March 2014

The complementary Water Icons and Color Palette includes carefully selected colors and iconic designations to represent the various types of water and to define water quality levels as well. The logic behind the choices can be defined as follows:

1. A universal "texture" symbol was chosen to symbolize water treatment
2. A color was chosen to most closely represent the kind/type of water it was set to represent
3. Icons, which include the chip of color or color in the internal object design, were created to compliment the color representation
4. The A, B, C, C*, D, E letter icons represent or designate the classes of water or water quality level
5. The CMYK, RGB, and HEX coordinates are provided to ensure color use accuracy and consistency in the development of water communication materials, such as education materials, infographics, technical white papers, marketing materials and videos.

We should start the industry standardization process with water associations and water and wastewater utilities, but certainly not stop there. Water and wastewater utilities face unprecedented fiscal issues, aging infrastructure, rising costs for service, the need for more training, expenses for regulatory and treatment controls, as well as weather-related challenges.

More often than not, the voice of the utility is not heard or represented in its various stakeholder communication channels – this can change. As utilities undergo quantum changes brought on by these challenges, efficient asset management is critical. Integrated management plans for the future must have a stronger emphasis on strategic communication. A solid communication portfolio and a standardized color palette should be positioned as utility assets.

Strategic communication, coupled with leadership support and sufficient resources, can help water and wastewater utilities achieve their mission, and it will designate a utility as a vital enterprise that's essential to thriving, sustainable communities. Strategic communication can also improve the public perceptions of the value of water and of the water utility.

It's been said, by water leaders, that the single greatest benefit to water infrastructure asset management is our ability to explain water. We can (and need to) improve the way we communicate about water and the value of water to society. Informed and educated stakeholders will think about, value and manage water in a different way, and that is a very good thing. A standardized communication effort will add value to our efforts in achieving this goal.

Engage in the Standardization Process

I invite your participation in using the model and the accompanying tools (conceptual model, water icons and the water color palette) to help you communicate about water. I also invite your comments and feedback. If you would like to send me a sample of how you used these tools, please send to donna@donnavincentroa.com. I will feature some of the water communication outputs on my blog Speaking Up About Water. Access the following link to download a zip file that includes high quality jpegs of the model, the color palette and the industry icons. http://donnavincentroa.com/roa-conceptual-model-for-water-communication/

About Donna

Donna is an internationally accredited business communicator, water communication expert, and strategic counsel to CEOs, scientists, engineers and professional communicators. She delivers value-based, industry-relevant communication and brand solutions and results to global water leaders at highest levels of business, government, associations, and international organizations. She frequently writes about water communication issues.

[8]

Utilize Social Media – don't block it. – Todd Danielson, P.E., BCEE

NOT TOO LONG AGO, social media tools were banned from many workplaces. However, the opposite should be encouraged, especially for utilities and water authorities. Effective use of social media improves a utility's effectiveness, customer service and it can reduce countless hours of work to answer complaints or response to bad publicity.

> "Social media is the ultimate canary in the coal mine." – *@JayBaer (via Twitter)*

Your organization may not typically use social media, but many of your customers use it regularly. Many utilities across the nation are in the midst of major capital improvement programs for aging infrastructure as well as responding to extreme weather conditions or gradually raising rates – all of these are potential fodder for complaints through social media.

And one complaint could spark the proverbial match for a firestorm of bad publicity.

For example, *The Plain Dealer* intermittently reports on Cleveland Water, a major regional water supplier. When an article about Cleveland Water is placed online, disparaging comments appear from their customers and certainly show the undercurrent of public sentiment, but virtually no one defends Cleveland Water. Even the supplier rarely defends itself in this case.

Utilities need to focus on the benefits of social media, such as how it can help gain public support and reduce customer complaints. At my utility, Avon Lake Regional Water, we used these new tools to inform the media and respond to concerns during a water crisis in early 2014. Making this type of effort helped us gain public support and allowed us to work on the issue at hand.

With a 50-degree drop in temperature and slight turbulence on Lake Erie, our company experienced a partial ice blockage of the intake screens for our water filtration plant. Small shards of ice, known as frazil, were entrained in the water column and allowed ice to accumulate on the intake screens approximately 20 feet below the water surface. As the intakes were blinding over and staff was unable to clear it, the primary water supply for approximately 200,000 people in northern Ohio was in jeopardy.

Obviously, this became a big story for the local media, so we decided that informing customers now would be better than damage control later. In addition to working with the media outlets, we began using Facebook to keep customers informed as to the status of the situation.

During the event, our Facebook followers grew from approximately 250 to 1,750; overall, we reached approximately 63,000 people during the event. Hundreds of people commented on the page with an overwhelming majority thanking us for all we did to get the water flowing again, but also for providing a flow of information during the crisis.

Figure 1: Excerpt from www.facebook.com/AvonLakeWater

By keeping customers informed about the situation and what we were doing to resolve the problem, our utility avoided negative customer opinions and gained customer support. When *The Plain Dealer* posted the article about our icing event, there were some disparaging comments, but several responses defended Avon Lake and squashed a lot of the bad sentiment. As an additional benefit, we shared the positive comments with staff to boost morale, and this even converted several staff members who were non-believers in the power of social media.

In addition to responding to crises, social media can be a great way to inform the public about issues. During the initial stages of a $2.7 billion sewer system improvement program, San Francisco Public Utilities launched a multimedia marketing campaign to sensitize the public to the sewer system under their feet. The campaign went viral – people were snapping selfies in front of advertisements with phrases like "No one deals with more crap than I do."

Figure 2: Photo from https://www.facebook.com/SFWater/photos

Many national news services and websites carried stories, such as this one from Fast Company (http://alwtr.us/PoopHumor). The campaign certainly had the desired effect of alerting customers about the investment, the importance and the benefits of the underground infrastructure.

Social media has the incredible ability to get information out to large populations in very short periods of time. Avon Lake learned from the frazil experience because, after an extended period of cold weather and snow, a significant thaw and rain event was predicted within our service area. This raised concerns about the potential for sewer surcharges and basement backups, along with overloaded basement sumps and other stormwater issues.

We used social media to alert customers to the potential issue and asked customers to take precautions. Luckily, the rain event did not reach critical levels, so surcharges and floodings did not happen. However, postings such as this help to reinforce that we are looking out for our customers.

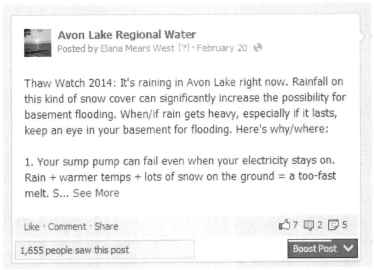

Figure 3: Excerpt from www.facebook.com/AvonLakeWater

Whether we're posting warnings about a potential crisis, giving advice to help customers to avoid major service issues or just providing fun information, utilities should employ social media to develop a relationship with the public and enhance our reputations as trusted experts who are looking out for our customers.

For instance, because our lakefront city experiences combined sewer overflows, we have always collected bacteria samples at the local beaches during the summers. Historically, these sampling results were provided to the recreation department which then determined whether the beach would be open or closed for the day. Last summer, we began posting the bacteria results on our Facebook page. Bacteria levels were closely related to rain events, and moderate rains could easily cause levels to rise above recommended maximum levels for recreational contact.

Beachgoers had previously been unaware about the bacteria levels, but, by providing the results on Facebook, we met several objectives. Our primary objective was to protect customers' health and safety, and a second objective rose from this because customers would no longer spent hours on the road to drive to a closed beach. Thanks to this information, we increased our Facebook following increased by 25 percent, and we educated customers to a key reason why we're separating combined sewers.

Utilities have historically been the "behind the scenes" silent service provider. However, we believe that should no longer be the case. When a utility needs to quickly get out important information, customers need to know by whom they are served – social media offers that immediate name recognition and customers remember who provided this helpful information.

This way, when potentially negative information is out in the media, customers who recognize the utility are less likely to see the details get misconstrued or taken out of context. One wastewater utility, the Northeast Ohio Regional Sewer District, used a public service social media campaign as one way to build name recognition. Their #WaitToTweet campaign, with supporting advertisements in bathroom stalls and above urinals at local sports arenas, is a local hit. The campaign subtly builds name recognition for the utility, while providing a memorable public service announcement.

Figure 4: http://neorsd.org/waittotweet.php

Social media's primary role is to build relationships, so a utility has the opportunity to directly connect with its customers. This allows two-way communication that can prevent minor issues from turning into major public relations events. Rather than relying on the traditional media to cover a story, utilities can create their own news and present it to interested customers.

Thanks to the exposure to such a wide audience, this can also create an interest where none existed before, such as Avon Lake's approach to the beachfront bacteria levels. Utilities must be willing to dedicate time to social media, even if the results cannot be immediately measured. At Avon Lake Regional Water, we've been investing in social media for years.

With positive feedback from the frazil event and our continued stream of information, we can start to measure the results in terms of customer satisfaction and public support. That's what all utilities need to aim for, so drop the firewall and start utilizing social media now.

About Todd

Todd Danielson, P.E., BCEE is the Chief Utilities Executive at Avon Lake Regional Water, a utility company based outside of Cleveland, Ohio. Prior to this, he worked at Virginia-based Loudoun Water for nearly 10 years, and Todd also has several years of experience as an environmental engineer.

[9]

Getting beyond dots on a map: Collaborative science and scenario planning in the Upper Gila River Watershed - Kelly Mott Lacroix & Dr. Sharon Megdal

> When you are dying of thirst it's too late to think about digging a well. – Japanese Proverb

WATER MANAGEMENT AND RESOURCE PLANNING are difficult tasks in the southwestern United States. One day, it's hot and dry; the next day, fields are flooded and roads are washed away.

The uses for our water are almost endless, and each drop, in some way, fuels our economy; no water means no fields, no copper, no cattle, no restaurants, no fishing and no trees. While use by one population does not necessarily mean water is unavailable to another, this valuable resource – at the price we are used to paying for it and the quality we are accustomed to – is increasingly scarce, especially in the more rural parts of the Southwest.

In the face of increasing scarcity and overlapping demands, how does a community decide how they will use their water? As the Japanese proverb implies, once you have run out of water, it is too late to think about where more will come from.

Planning for resilient futures in our communities and watersheds requires many things:
1. an understanding of what resources a community currently has,
2. how water law and policy impact the use of these resources,
3. how those resources have changed over time,
4. what resources might look like in the future, and
5. most importantly, how residents want to use these resources to shape their community.

Water management planning at the watershed scale offers the opportunity to consider a wide range of inter-related resources and issues in a unified way. It enables a community to look comprehensively at human and environmental resources throughout a hydrological system. Traditional land use and city planning do not take this comprehensive approach.

Watershed planning with Gila River communities in Arizona

In October 2012, the Water Resources Research Center (WRRC) at the University of Arizona embarked on a project to work with the Upper Gila River Watershed communities in southeastern Arizona to enhance understanding of and planning for their watershed. The two-year project is funded by the Desert Landscape Conservation Cooperative, and it's a partnership between the WRRC, Arizona Cooperative Extension and the Gila Watershed Partnership.

Our ultimate goal was – and remains – to get beyond just displaying dots on a map. We wanted to collaborate with community members from the beginning of the project, whether it was shaping everything from the types of data we collected or building scenarios for the future around their interests and concerns.

Past efforts at understanding the watershed were mostly conducted by researchers or state agencies, and they generated interesting and insightful final reports. Unfortunately, most of those reports collected a lot of dust because they were unused by the community – and those reports were specifically written for their planning and management purposes.

Our challenge, therefore, was to design a process where we could do the necessary heavy lifting of research and report creation. At the same time, we wanted to involve the community in co-production from the outset, so that the results of the project were useful and usable to them.

We had a critical question to answer: in a rural Arizona community, where funding for planning, management, and natural resources research is limited, how do we balance the need for community involvement with their time and resource constraints?

Phase 1: Determining Key Values and Data for a Valid Assessment

The first phase of the project was the development of a baseline assessment of watershed resources, which collects and displays key information identified by stakeholders as well as the current conditions of natural and water resources. The watershed assessment also shows how resources have changed over time and identifies some of the challenges the watershed faces in the future based on current conditions; this process helped with the creation of the *Atlas of the Upper Gila River Watershed*.

Using the expertise from the Gila Watershed Partnership (our local partner), we began by forming a Watershed Assessment Workgroup – simply called the Workgroup – to assist in defining the parameters of the baseline assessment and identify data sources. One of the first concerns raised by the Workgroup was the need to coordinate the collection and storage of data.

This concern led us to convene a Science Coordination Workshop for the watershed. The two-day long workshop brought together scientists and land managers – from different areas throughout the watershed – to share data and information, identify opportunities to coordinate on projects, and contribute to the goals of the larger watershed assessment project. The meeting resulted in identification important data gaps and opportunities for coordination to address these gaps.

One of the biggest data gaps identified during the workshop was the need to capture local knowledge, history, and understanding. While a number of books have been written on different aspects of the watershed, such as *Gila – River of the Southwest* (1951) by Edwin Corle, participants in the workshop felt that much of the local insight and understanding was disappearing as long-time residents moved out of the valley or passed away.

To begin to fill this gap, we hosted a four-hour long Shared History Timeline exercise held in conjunction with a monthly Gila Watershed Partnership meeting. We invited the community at large to participate in the meeting, and we were pleased with the large attendance at the event, which included a number of ranchers and farmers that spanned generations. For this "shared history" exercise, participants were asked to help combine their personal knowledge (along with empirical data) to create a timeline of the history of the Upper Gila River Watershed.

The results of this shared history exercise became a critical piece of the watershed assessment. This event helped to reflect the collective experiences of local residents in adapting to the unpredictable cycles of drought and flooding in the watershed; it provided an extended perspective over time that could provide insight for any events in the future. The paper document created was also turned into an interactive online timeline for the Gila Watershed Partnership to use into the future: consisting of video recordings of stories from participants at the meeting and historical photos, the timeline is available on the WRRC and GWP webpages at http://wrrc.arizona.edu/Upper-Gila-Timeline.

The Creation of Geospatial Information

With the data collected through the Workgroup, the Science Coordination Workshop and the Shared History Exercise, the WRRC built a geospatial database that contains 62 unique spatial datasets, ranging from soil erodibility to instream flow water rights. These datasets were then collected into thematic maps, which would provide both context for conversations about watershed management and how the watershed has changed.

Examples of maps within the Atlas are:
1. areas with special management such as national parks, wilderness areas, and critical habitat,
2. groundwater conditions, including depth to groundwater and well yields, and
3. change in water demand over time.

Because part of the goal of the project was to assess change in the watershed, we also built an Automated Geospatial Watershed Assessment (AGWA) model to show how key functions, such as runoff and sediment movement, have changed over time. The AGWA model was originally developed at the University of Arizona with the U.S. Department of Agriculture's Agricultural Research Service and the U.S. Environmental Protection Agency. It has been used in a number of watersheds in the desert southwest, and is able to calculate sediment yield and runoff when given data for topography, soil, land cover, and precipitation.

Creating an Atlas for the watershed has served multiple purposes, such as bringing stakeholders together to discuss data first, which has allowed a diverse set of people to come together – in a non-adversarial way – at the initiation of what will be a multi-year watershed planning effort. The Atlas has also ensured that everyone starts on the same page, with a common understanding of both watershed history and resources.

Furthermore, the project team has built an important bridge between the community (that will use the information) and up-to-date scientific information (by having the team spend time in the community and learning about their concerns or ideas). These community discussions have been critical in the co-production of the watershed assessment, as various drafts have been reviewed at stakeholder meetings, then they have been continually revised in order to reflect the views of the residents.

Remember that creating a watershed assessment is not a watershed plan; however, it is the important first step toward understanding existing resources and making informed decisions about their management. The watershed assessment then paves the way for development of scenarios for the future of the watershed, which is the second phase of the project.

Phase 2: Utilize the data to prepare for various situations

Scenario planning is a strategic process designed to help people think systematically and creatively about the future, especially when provides insight into the forces that shape the watershed. Understanding these forces helps reveal the implications if the group follows the status quo, it explores what might happen in the future, and highlights the various ways to take action.

Scenario planning helps people make decisions today that could result in the type of future that they want. Using the common understanding from the watershed assessment, the WRRC is working with the community to answer two critical questions: what is causing change in the watershed *and* what are the critical uncertainties that natural systems, the economy, and social-political life in the area face?

In answering these questions, similar to our approach in the first phase of the project, we have tried to strike a balance between community involvement and stakeholder fatigue. We wanted enough community input to make the results relevant and useful, but not so much that the process becomes overly onerous for busy participants. Traditionally, scenario planning requires a significant time commitment on the part of the participants, including several two-day or three-day workshops.

During conversations with our Workgroup, we learned that there is considerable interest in developing scenarios for future planning. Participants, however, almost universally expressed that they could not set aside large chunks of time necessary for a traditional scenario-planning process. So we opted to conduct a single, day-long workshop.

It was very productive and well received, although intense and rather demanding, and this condensed meeting was still difficult for some stakeholders to attend. This affected farmers and ranchers who could not take a day away from spring planting, so we arranged a separate, mini-workshop in the evening to gather critical input from them.

The remainder of this phase involved monthly hour-long meetings with the Watershed Assessment Workgroup. We used individual meetings with participants – especially those who would use the results for future planning and management – to ensure that the final project product met their needs. We also incorporated the use of online tools, like Survey Monkey, to broaden participation by people who could not attend workshops or monthly meetings.

The end product of the project is a set of scenarios, created with data from the watershed assessment and the ongoing conversations with stakeholders. These scenarios describe the consequences of possible alternative future conditions in the watershed, and they explain legitimate ways that these communities can use the information for planning and managing their resources in the coming years.

Communities need to take action today rather than take a chance in the future

Uncertainty in water resources for communities like the Upper Gila River Watershed is, well, a certainty. There is very little that can be done to guarantee the amount of groundwater and surface water available in the future, but there is much a community can do to manage what they have today. This management requires a common understanding of the existing resources as well as a shared commitment – between all users of water – to collaboratively manage their resources through times of uncertainty.

Universities, particularly those with Cooperative Extension units like the University of Arizona, have a reputation for their ability to serve as neutral and trusted mediators. They have an important role to play in enabling communities, whether it is by helping them assemble and analyze information or through exercises like shared-histories and scenario planning; as described previously, this creates a common understanding across different groups of water users.

Iterative interaction, between academia and communities that need the science and planning support, can enable management that supports resilient communities and watersheds. And this is crucial when facing climate change and other pressures on our most precious resource: water.

About the Authors

Kelly Mott Lacroix is a research analyst for The University of Arizona's Water Resources Research Center in its Water Research and Planning Innovations for Dryland Systems (RAPIDS) program. She has a master's degree in Environmental and Healthy Cities Planning from the University of Arizona.

Sharon B. Megdal is Director of The University of Arizona Water Resources Research Center (WRRC), and she is a C.W. and Modene Neely Endowed Professor in the College of Agriculture and Life Sciences. Dr. Megdal has served on numerous Arizona boards and commissions, including the Arizona Corporation Commission, the State Transportation Board and the Arizona Medical Board. She has a Ph.D. in Economics from Princeton University.

[10]

Water Clusters and the Acceleration of Innovation – Jon Grant

THE CHALLENGES FACING THE WATER SECTOR are too vast for any one jurisdiction to come up with all of the answers, so regional clusters are forming to reap economic benefits. However, they realize that cooperation and collaboration that sector is the only way that our industry has a chance to conquer these challenges. Much like a pack of wolves hunting animals much larger than themselves, water sector professionals need to come together – through online platforms and water clusters – so we can tackle problems that are bigger than all of us combined.

WHAT ARE WATER CLUSTERS AND WHAT ROLE DO THEY PLAY IN OUR SECTOR?

Water issues are global, but its local impact or region-specific problems may require different solutions to face similar problems. For example, Israel pioneered drip irrigation out of necessity. Ontario, by contrast, would not have needed to innovate efficient irrigation (until much later) because it sits on the shores of 20% of the world's fresh surface water supplies. The same type of issues can be approached with a variety of tactics, and water clusters are used to focus on the right ones.

Generally defined as a regional group working towards solving at least one water issue throughout the development stages, a water cluster usually consist of members from universities, multinational corporations and everything in between. The major focus is eliminating the risk in water-related investments and adopting better solutions in the sector. In the past 5 years, groups have seen the value in creating a central hub of information and championing growth in the sector.

There is a lot of research on clusters as economic industry groups, most notably from Harvard's Michael Porter.[1] He has described these clusters at length in other industries, but this is a relatively new phenomenon as it pertains to the water sector.

Clusters create advantages across the spectrum for those involved and turn into an exercise; it's like a survival of the fittest, but with expert support. In line with the metaphor from before, if that pack in the wild suddenly received expert mentoring on training and skills that improved its ability to hunt, then what would happen? The pack would increase its size, boost its chance of success in finding enough food and to have the ability to effectively raise the next generation of wolves.

This is why water clusters are gaining popularity, especially with the ever-growing concerns about water scarcity. Due to the multiple uses of water, each cluster can understand its needs within the broader context, whether they focus on quality, quantity, stormwater

---

[1] http://www.economist.com/node/14292202

and/or a number of other areas; this will likely mean they have a diverse group of representatives.

For example, in looking at the structure of the water sector in Ontario, there are at least 9 technology sectors within the provincial water sector (most likely 12 to 15). Manufacturing ultraviolet lamps for disinfection is quite different than integrating entire treatment systems into a plant, but the interdependence of these niches add some complexity to the sector itself.

Water clusters & critical research

Water clusters focus on a variety of things: promoting research, testing, validation, entry into the market, and growth through export. The details of how that is carried out may change, but the main goal is to foster innovative solutions.

University research is a foundation of any high-tech sector, and it's no different in the water industry. Perhaps it's even more essential, especially with rising issues that surround the microscopic contaminants and the multiple variables that go into treating water. Students are encouraged to push the bounds of their fields and develop new technologies, which can turn into future water solutions.

In looking at research, clusters promote technology that can be commercialized: its one thing to have a novel and interesting technology, but it's a completely different to have a product that fills a demand. Therefore, there are generally a lot of representatives from small, early-stage companies with new technologies that most likely need to be proven in real-life situations.

Through a series of testing by the companies (and potential end users), the valuable data and feedback enables these companies to focus on innovation without risk to people or the environment; the tests are based on legitimate situations, but aren't directly connected to the water supply. Regulators also have their say by setting tough (yet sensible) regulations, which encourage these researchers, end users and companies to innovate within the legal parameters.

The end goal for the cluster is to get innovative products to the marketplace, which is a tough task in nearly any sector. Unfortunately, the general rule of thumb in the water sector is that any technology will take about 10 years from initial creation of IP to market acceptance. With the data obtained during these experimental tests created by water clusters, the risk factors are reduced and the adoption process can be sped up through common understanding of the technology.

It's good to see that the U.S. Environmental Protection Agency has been hosting meetings between North American clusters; at the time of publication, there are at least 15 of these clusters. Each has a different approach on how to best accomplish the goals of solving water issues, and each cluster brings a number of strengths in different areas.

For example, understanding efficiency in the Southwestern U.S. will likely differ from the strong treatment technologies being developed in places like Cincinnati, Ohio. Plus, these clusters will have various structures of funding, since some are government-funded groups while others are funded by memberships or board seats.

How water clusters strengthen innovation and commercialization

There is a concept of the factor endowment model, which states that countries or groups are strong because they are in an environment that provides the right tools for success. In other words, necessity is the mother of innovation.

For example, Israel has very little water and therefore developed a technique for irrigation to maximize efficiency, which became part of its economic strategy. Areas that have a competitive advantage in producing one type of good should trade for goods they need or want from an area that produces those goods at a competitive advantage. In this case, Israel exported its drip irrigation technology to other countries while importing various technologies that would provide basic functions in other industries.

Water clusters create a positive feedback loop for universities in three ways: through research, more funding and a pathway for graduates. Building intellectual capital in water encourages researchers to push technology barriers even further, especially if the end result has a commercial focus. Then, by bringing industry into the fold, it generates an increase in funding for this particular research. And when the industry has an eye on the talent being produced by the university, those new graduates are much more likely to be hired.

For early-stage companies, they benefit from being part of a water cluster through three avenues: support, connections, and awareness. Other cluster groups, incubators, industry associations, and other companies can support these smaller companies through strategic partnerships; as such, these clusters can provide awareness and connections to the right resources at the right time.

High-growth companies are able to accelerate their growth by using the connections of the cluster to help gain vital contacts into markets they wish to enter. Programs created by the clusters enable them to learn and enhance vital business skills, which allows them to grow their businesses and market their products.

And large companies can several benefits if they join a water cluster. They are able to access groundbreaking research through partnering with universities, while being able to partner with (or acquire) smaller companies with new technologies to round out their core offerings.

Now, coordinating these groups within clusters can be tricky because the sector, as many insiders will agree, is fractured. Hubs at the center of these clusters enable the speed to market – from basic research to adoption – to be reduced.

They approach it through different methods, but, with the same goal in mind, they all help create solutions for end users so we can solve water problems. With that in mind, let's examine the origins and current progress of a leading water cluster.

Ontario is a Key North American Water Cluster

How did it come to be that a region on the shores of the largest amount of surface water would grow into a water technology region? This is what has happened to make Ontario one of the leading water clusters on the planet. After a lengthy time spent speculating on this, my educated guess is this rose out of necessity.

Ontario has more water than it knows what to do with, so what gives? There are two things at play in this case: water quality and location. Even though the majority of the population lives around two downstream sources, Lake Erie and Lake Ontario, pollution is the driving factor here.

Toronto, and its large metropolitan area, is downstream from the cities of Milwaukee, Detroit, Cleveland and other areas historically known for industrial manufacturing. This production, unfortunately, sent extremely polluted water back into the Great Lakes; even the Cuyahoga River in Cleveland has been on fire multiple times due to the pollution levels, most recently in 1969.[2]

2

http://www.cleveland.com/science/index.ssf/2009/06/cuyahoga_river_fire_40_years_a.html

As a matter of survival and necessity, Toronto got really good at cleaning out water through innovative techniques; it was one of the first cities in North America to use chlorine to treat its water.[3]

Filtration also became a high priority and a topic of importance; even before economic clusters began, universities and research institutes were on the cutting edge of research. The story of Zenon, the company that revolutionized membranes and drinking water, is a true Ontario water story when it comes to filtration. The technology came from university lead research and this developed into a global leading company.

Other companies, such as Trojan UV and the Pressure Pipe Inspection Company (PPIC), revolutionized their respective markets. Trojan provided municipalities with the ability to disinfect their water supplies without adding chemicals, and PPIC enabled municipalities to understand their buried infrastructure and avoid major catastrophes from large-diameter pipe breaks.

These companies – Zenon, Trojan and PPIC – form the basis of the technology cluster. Even though all three have been sold to larger companies, they still maintain a core level of personnel in the water cluster.

And this is just a sliver of what's going on because, on the research front, there are over 100 water-related programs in Ontario alone. The research and early stage expertise is being harnessed on a number of different organizations, like the Southern Ontario Water Consortium (SOWC) and the Ontario Centres of Excellence (OCE), so university-driven research can flourish and produce incredible new technologies in water.

3

https://www1.toronto.ca/wps/portal/contentonly?vgnextoid=54a807ceb6f8e310VgnVCM10000071d60f89RCRD&vgnextchannel=6d1409f8e0c7f310VgnVCM10000071d60f89RCRD

These organizations also aim to help innovation without too much governmental interference; Ontario has been noted to perform poorly in business innovation.[4] Due to this, the Provincial Government wanted to improve those concerns when it comes to a valuable resource that's important for the province as a whole. So the Water Opportunities Act of 2010[5] was passed, and it gave a clear mandate for the development of innovation in the sector.

WaterTAP Should Boost Ontario's Water Sector

The Water Technology Acceleration Project – typically known as WaterTAP – was created as part of the 2010 act. The project has a mandate to grow the water sector and act as a hub of information, and it accomplishes that role by working with small and medium-sized businesses.

Developing programs to help these businesses improve their basic marketing and strategic functions is an important part of the project, but creating connections to resources is a vital role – WaterTAP is able to introduce companies to the right people at the right time. In addition, the project has become a mouthpiece for the local water sector on the global stage, it is part of the regional water cluster, and celebrates the success of Ontario's research and solutions.

Like its counterparts in Germany, Holland, Singapore, Israel and the U.S., WaterTAP is trying to improve outcomes with regards to municipal and industrial water. No one jurisdiction is going to solve all of

---

4

http://www.stic-csti.ca/eic/site/stic-csti.nsf/eng/00076.html#backgrounder

5

http://www.e-laws.gov.on.ca/html/source/statutes/english/2010/elaws_src_s10019_e.htm

the issues because all water issues are inherently local; Ontario won't heavily focus on conservation technology in comparison to places that need to produce highly-efficient solutions like Israel.

IT'S TIME THAT OUR SECTOR FORMS A PACK – AND, NO, NOT A PACT

Water clusters are accelerating the creation and adoption of new technology. Through the knowledge gained and the networks they create, our fragmented world of water can start to collaborate in one cohesive unit. The transfer of data and information through the global network of clusters will enable faster adoption of innovative technology, which will help conserve more water and allow us to make better use of our resources.

We need to come together and look out for solutions – then hunt them down like a pack of wolves.

ABOUT JON

Jonathan Grant is the Manager of Research for Toronto-based WaterTAP Ontario. He has a bachelor's degree from the University of Cincinnati with a double major in Finance and International Business, and he later continued his studies at the London School of Economics for an MSc in International Political Economy. Because water issues are clearly connected to political and economic issues, Jon also earned an LLM in Water Governance and Conflict Resolution.

[11]

Accelerating Innovation – McGee Young

Fewer than one hundred startups, out of nearly 300,000, identify "water" as their primary market. Given the magnitude of water-related challenges that span vast markets of customers, why aren't more entrepreneurs flocking to water-related opportunities? Would more programs or resources improve the rate of water entrepreneurship?

Experts frequently cite three factors for the lack of water entrepreneurship (Fishman 2012):

1. water is cheap, relative to other commodities,
2. water is a public resource that's managed by bureaucratic methods, so innovative thinking can be limited, and
3. water is boring because people take it for granted and rarely produces celebrities.

While the water sector is not particularly popular among startups, there is still considerable innovation happening, although it's "hidden" because large incumbent firms still dominate the industry; many new technologies come from companies with large R&D budgets. Only in the last few years have organizations emerged to try to elicit more innovation from new startups to meet the challenge of future water problems.

> Given these challenges, what can a water accelerator deliver? Recognizing the limitations of the sector, what can we hope for from water startups, and what should we expect?

The Birth of a Global Water Center

Milwaukee, Wisconsin: former home of the nation's largest breweries, some of its most productive tanneries and an assortment of heavy industries that made the city part of one of the most productive regions in the world. Industrial-era supply chains required close proximity, so around this hub of anchor industries grew a network of smaller suppliers that provided pumps, valves, meters and other equipment needed to keep the factories running.

But even as most of the major manufacturers to shut down or scale back operations, the globalization of the supply chain meant that niche companies like Badger Meter could continue to thrive in Milwaukee. This meant that, without really trying, Milwaukee ended up with more Fortune 500 company headquarters per capita than any other place in the United States.

Many of these companies have roots in the water industry, whether as large consumers of water (like the Miller Brewing Company) or as producers of water products (like the hot water heater manufacturer AO Smith). And the city's wastewater treatment agency, the Milwaukee Metropolitan Sewerage District, has gaining a reputation as a national leader in innovative wastewater treatment strategies. It takes innovation to find ways to reduce the amount of raw sewage dumped into the region's rivers and Lake Michigan.

With all this potential, Milwaukee's corporate and civic leaders established the Water Council, a membership-driven trade group comprising businesses, universities, non-profits, and government agencies, in 2007. After five years of growth, the Water Council proposed an ambitious plan to renovate a 100,000 square-foot office building next to an abandoned rail yard. This was to help convince companies from around the world to relocate to a water technology office park. Local and state political leaders contributed tens of millions of dollars to a catalytic investment, and the first building appropriately named the Global Water Center.

The Global Water Center (GWC) opened its doors in September 2013. Many of the largest water companies in the world, from Veolia to Sloan Valve, joined Badger Meter and AO Smith in leasing space in the GWC. Rexnord announced that its entire executive leadership team, including the CEO, would relocate there. The University of Wisconsin-Milwaukee's School of Freshwater Sciences leased an entire floor for offices, classrooms, and laboratories.

And two floors were occupied by a select group of companies with unfamiliar names: Vegetal ID, NewWorks, H2Oscore (my company), Microbe Detectives, and Rice Technologies. These companies constituted the startups in the Global Freshwater Seed Accelerator and were joined by other young companies, such as Hanging Gardens, Alga Bionics, Pave Drain, and Xela Innovations.

The Water Council made a huge bet: provide office space and access to industry leaders for young companies, and they will grow and mature into the leading innovators of the next generation.

The Global Freshwater Seed Accelerator

There are two main types of programs that support new businesses: the seed accelerator and the business incubator.

The *seed accelerator* model typically follows a specific formula where, once or twice a year, there's a small equity investment into a small portfolio of companies, and the companies receive intensive business development guidance and mentoring. After four months of progress, the hosting organization holds a high-profile event to spur investment into these new companies.

By contrast, a traditional *business incubator* will typically provide office space for a longer period of time, but will not make an investment into the company. It also will not provide a forum for the incubated companies to pitch to investors.

The first year of the Global Freshwater Seed Accelerator provided companies with a hybridized model: all of the startups were provided office space in the Global Water Center, but instead of an equity investment, the new companies received $50,000 grants from the State of Wisconsin. The typical four-month session was extended to a full year, and, instead of a demo day, the startups were recognized at a "graduation" ceremony that was officiated by the Governor of Wisconsin.

Even with appropriate funding, startup accelerators face two challenges: the job to be done with mentoring, advising, motivating, and advertising, and the startups themselves. Startups are prone to failure and they fail for a wide variety of reasons. Aside from Y Combinator and TechStars, very few accelerators have been able to consistently launch high-growth startups.

To solve the first challenge, the Global Freshwater Seed Accelerator (GFSA) contracted with the University of Wisconsin-Whitewater's Institute for Water Business to provide business development services for the startups. Twice weekly, faculty from the University would lead the teams through mentoring sessions and brought in guest speakers to share insights into the various challenges that were facing new companies.

The heterogeneity of the startups in the program complicates the work of the Accelerator staff; this is the second challenge with seed accelerators. Not all GFSA startups are aimed at high growth or large markets – some are focused on hardware, others on software, and a handful are building products that combine the two. Some of the startups have been operational for a year or more, while others were mere prototypes of products that lacked any commercial validation before entering the program. There is very little common ground around which to build a startup curriculum or establish metrics of success; even as a unique challenge to the GFSA's model, it was still a reflection of the diversity of the water industry.

One other concern is the State of Wisconsin's political climate, which (like most states) revolves around the health of the job market. Because each of the startups receives a $50,000 grant from the Wisconsin Economic Development Corporation, the key metric followed most closely is the number of jobs created from the program. While employment amongst the companies in the Accelerator more than doubled over the course of the first year, significant questions remain whether job creation is even a good metric to evaluate the success of the program.

Can We Prime the Water Innovation Pump?

Rich Meeusen, the voluble CEO of Badger Meter, identifies the café on the First Floor of the Global Water Center as the most important "center" for innovation in the building. The argument he puts forth starts from the assumption that when entrepreneurs, researchers, and experienced business professionals share ideas, new insights into solving old problems begin to emerge.

Aside from the afternoon coffee klatch, more formal and substantive events and organizations also spur innovation at the Global Water Center, such as:

- a hardware hackathon where industrial designers, software developers, and business professionals met for a weekend to design and build new products
- the NSF-funded Industry-University Cooperative Research Program (IUCRC) holds meetings and workshops in the conference space to facilitate the commercialization of research emerging out the city's three main research universities, and
- weekly formal presentations of work-in-progress help build familiarity and raise awareness of the variety of types of research and development projects occurring within the building.

In short, a stunning range of new ideas, technologies, and solutions are being developed within the walls of the Global Water Center. It's demonstrated by the second round of the Global Freshwater Seed Accelerator, which included a dizzying array of applications from around the world. Innovation hardly seems to be a barrier to entry in the water sector because there are so many ways in which our water systems are outdated, poorly performing, or fail to solve even the most basic problems of supply and treatment, so even incremental improvements on the status quo present incredible opportunities.

While the café at the Global Water Center may further spark creative energy, it does not address the more underlying set of challenges that constrain innovation in the water sector. Ironically, the very system that makes the Global Water Center possible – huge incumbent companies underwriting innovation – is what stands in the way of even more innovation. It seems that the challenge to startups is not how to generate more innovative ideas, products, or solutions, but rather how to disrupt the system that prevents the diffusion of these innovations into the market.

The Dilemma of Innovation

In his classic treatise on innovation, Clayton Christensen showed how incumbent firms failed to notice or adapt to technological changes. From disk drives to steel, history has seen incumbent firms displaced by innovators leveraging new technologies, although the speed of innovation varies widely. The rapid evolution of personal computing meant that the disk drive market was rapidly disrupted several times within a single generation, while the steel industry withstood disruption for several generations, even with the development of new technologies.

The water industry bears much closer resemblance to steel than to disk drives: products are purchased infrequently at a relatively high price, customer relationships are well developed (and are highly personal), and significant capital is required to enter the market. Consequently, it is no surprise that most innovation to reach the market has come from incumbent firms, whether they deal with water treatment facilities or Automated Metering Infrastructure.

When we speak of innovation in the water industry, it is clear that the status quo is far more common than disruptive innovation from upstart companies, which harness the newer technologies to displace incumbents. Although, with large cash reserves, many incumbents can afford to buy small upstart companies and incorporate their technologies into existing products.

The question to be raised is not whether the water industry can innovate (or even if it does innovate), but whether we should seek to promote and support sustaining innovation – using those products that add value to existing companies – or disruptive innovation, where new companies take market share away from incumbents. If the latter, is it possible for such innovation to occur within an environment that is largely managed by incumbent firms?

This is a question that has yet to be addressed at the Global Water Center. For now, existing companies and startups work alongside each other in relative harmony. But, as the task of the accelerator shifts from providing programmatic support to growing new companies, the ambiguity in the purpose of the Center will appear in greater relief.

Conclusion

White boards cover the walls of my office in the Global Water Center; on the corner of one board sits a baby giant panda. The panda represents an inside joke among our team after the first few months in the GWC: as a first of its kind building, the GWC drew visitors from around the world, and people on a tour would visit the startup floor to stick their heads in our office to gawk. So, after a while, we began to feel like the giant panda exhibit at the zoo because our work was interesting to look at, but no one wanted to get too close.

There is a duality to startup life in a place like the Global Water Center: legitimacy is instantly conferred by virtue of a place in the Accelerator – and office space in a stunning new building – but there are also frequent reminders that you are a threat to the status quo. Each one of the first-year participants in the Accelerator struggled to get pilot projects started.

After four months, I made the tough decision to shut down the servers that powered H2Oscore. Despite five municipal utility contracts serving nearly a million customers, it became clear to me that the service – which we had envisioned for consumers – required a market that simply didn't exist yet; the necessary pieces simply wouldn't align in a time frame that made sense to continue.

We will soon see the water industry become more disruptively innovative. In all likelihood, the calamitous effects of climate change will push the absolute need for fundamental structural change. In the meantime, we are likely to see innovation that primarily builds value on existing products and services, while the incumbent firms continue their dominance over the marketplace.

About McGee

McGee Young is the Entrepreneur Faculty Fellow at Marquette University, and he is founder of two Milwaukee-based ventures, H2Oscore and MeterHero. After receiving tenure from Marquette, McGee decided to apply his research to solving real problems, such as creating applications to help homeowners conserve water and save money. He has a Ph.D. in Political Science from Syracuse University.

[12]

Technology, Utilities and Innovation - from "What if?" to "Here's how" – Jim Ginley

WE OFTEN HEAR IN OUR SECTOR that necessity is the mother of invention.

Well, given the pressures facing utilities – stricter regulatory demands, aging infrastructure, scarcer resources and nothing's getting cheaper – it is fair to say that there's a whole lot of "necessity" to inspire new solutions or inventions to address the issues we face.

Yet, utilities worldwide continue to struggle to find these solutions, while innovative ideas from laboratories, think tanks and technology clusters are also going unnoticed. But collaboration is bringing more people together to find, share and develop technologies for the future of water.

It all starts with "What if…"

At the core of technological innovation, we blend a mix of imagination and curiosity. Then we add heavy doses of scientific and engineering capability, and, of course, funding. Think of the possibilities:

1. What if you could harness the kinetic energy and hydraulic head in a segment of your distribution system and turn that into electricity to either power to a pump station or dump it into the grid – and have all the capex bundled into a 20-year power purchase agreement?
2. What if you could add some "special sand" to your stormwater runoff ditches – to virtually eliminate discharge permit violations – and end up with an attractive landscaping feature that would last at least 5 to 10 years?
3. What if you could reduce your biosolids handling and transportation costs by 20 to 25 percent *while* increasing your biogas production by 25 to 30 percent – and simply by mimicking a technique from the ethanol production business by "blowing up" your sludge?
4. And what if you could effectively increase the reflectivity inside a UV tube, used for disinfection of drinking water, to afford an 80 to 90 percent reduction in your electrical costs *while* greatly reducing your operational footprint at the same time?

Because these ideas started with a heavy dose of curiosity, they are way beyond the "what if" stage. Now these technological features and projects have moved from idea to development, and some are currently delivering real results. The one thing they have in common is that they have evolved while spending some time in something called a "TAG."

So, what is "TAG"?

A few years ago, a small group of engineers and scientists in the UK were facing a nagging dilemma. Utility clients were demanding that they find game-changing ways to lower expenses, reduce upfront costs, increase safety, and improve the quality of the water they provided.

Sound familiar?

Well, the team contrived a solution that they dubbed as "TAG", which stood for the Technology Approval Group. It's a neatly organized – they are British, after all – and highly interactive forum, which is to accelerate the uptake of innovative, game-changing (if not "disruptive") technology into the commercial water sector. This is achieved by orchestrating the interaction of technology companies, utilities and financial investors.

It started with a dedicated group of utilities, some of whom were known throughout the UK or the world: Thames Water, Severn Trent Water, Scottish Water, Dŵr Cymru Welsh Water, Anglian Water, Wessex Water, Southern Water, and Northumbrian Water. The group began to talk about what they needed, what challenges they faced, and what problems they needed to solve – all to find a pathway for those new, different and innovative pieces of technology.

At the end of the discussion, the group created a list of potential technology companies that could help with these problems; through a voting process, five technology companies were given a chance to pitch to a captive, motivated and vested audience. And that's the story of TAG.

Really... it's just that simple? Yup.

Through a vetting and informational process, this TAG concept is occurring in places all over the world. From Australia to Singapore, from South Africa to Brazil, interested utilities are connecting with emerging technology companies that provide cost-effective solutions.

More than 450 new technologies have been presented and more than $650 million has been invested – when that much commitment to innovation exists, it makes you wonder why the water sector didn't start this earlier. As more TAG meetings are held, more information is gathered and more technology companies are moving down the path to entering the market. The process seems to be working smoothly, possibly due to the lessons about the dynamics of the water sector's attempt to find and embrace new technology. However, two key points stand out: utilities are playing a bigger role in the innovation process, and they are collaborating in the innovation effort, especially in research.

Utilities playing a role in innovation

By getting involved in the entire process, utilities are playing a bigger role in finding the solutions *with* the technology companies. It is taking place at regularly scheduled and highly-orchestrated forums, such as TAG, or in foundation-driven committees and research groups, such as the Water Research Foundation (WRF) and the Water Environment Research Foundation (WERF); innovation gets better when utilities get involved from the start.

A prime example of this effort is the Los Angeles-based Metropolitan Water District of Southern California (MWD), one of the largest providers of drinking water in the world. The MWD has hosted a regularly-scheduled forum every quarter since April 2013, and the system is also heavily involved with the process of fostering innovation through its leading role in an industry network called H2O Tech Connect.

H2O Tech Connect is a smart connection and matchmaking network that connects people throughout the water industry, such as entrepreneurs, university innovators, investors and service providers.

The MWD identified a handful of up-and-coming technology companies from within the H2O Tech Connect network, and one was chosen – Los Angeles-based GreenTech Motors – to make a presentation.

The company has developed a highly-efficient electric motor that can be used in commercial and industrial locations, including water sector utilities. This technology has the potential to help meet the escalating demand for cleaner, more sustainable electric motors, which would provide energy efficiency and lower costs as they reduce greenhouse gas emissions. Because of the collaboration between the metro water system and its networking partners, the fledgling technology company was given a unique chance to present to a room full of water experts and receive substantial feedback that will assist in preparing the motor for the marketplace – and this all happened in one meeting.

Utilities are working better with one another

Another example of collaboration comes from a place that is accustomed to inspiring utilities to work together, the Denver-based Water Research Foundation. For many years, WRF has run a collaborative research project program, whereby utilities identify the topic and bring their own research funds to the table to help get the work done. The WRF originally shied away from assisting with research beyond the basic proof of concept and from doing actual demonstrations in a trial setup at a utility.

Well, that has changed, partially due to WRF's subscribers asking for more demonstration and/or trial projects and also due to WRF's partnership with Isle Utilities.

For example, when a new UV water treatment technology was presented at a forum hosted by the Metropolitan Water District of Southern California, several utility companies were interested and wanted to arrange a trial setup to test the technology. The WRF found that other utilities, which were not present at the forum, were also interested in treatment technology of this kind. Within a few months of the presentation, the concept was pulled together under the auspices of the WRF. A third-party consultant/principal investigator was also assigned to the project, and utilities from California, Colorado, Texas, and New York were now part of the overall project. As of the date of this publication, two of the California utilities – East Bay MUD and Santa Clara Valley Water District – are physically hosting the demonstration.

It's quite amazing that this innovative high-tech company will have the benefit of a highly-regarded research project as the basis for its claims, and all WRF's subscribing utilities can learn from the results of this project. Even more amazing is how a 20-minute presentation is leading to a potentially game-changing body of information that could help the entire industry.

It's all about collaboration and innovation

Innovation – and the game-changing breakthroughs that will provide utilities the solutions they need for today's and tomorrow's problems – is not the sole domain of the technology innovators. Rather, it is a shared and collaborative effort... and it is growing (and morphing) rapidly.

The water sector utilities themselves are playing a larger and more significant role in redefining how it happens. Some are doing it practically on their own, but more and more are finding ways to work together through teamwork, partnerships, committees and creative networking.

With a highly-interactive exchange of information and interest that connects utilities and high-tech companies, we can expect that more utilities around the world will start to walk down that path toward innovation.

About Jim

Jim Ginley is a Senior Consultant at Isle Inc., a technology consultancy that accelerates the uptake of innovative and game-changing technology by cleantech sector agencies worldwide. Ginley has more than 27 years' experience in the municipal water sector, with nearly 17 years on the staff of the American Water Works Association, as well as being a regulatory compliance specialist and utility management consultant. Ginley is a member of AWWA and WEF and a frequent presenter and session facilitator at ACE, WEFTEC, and the Utility Management Conference. Ginley is a graduate of Kenyon College and Duke University and resides in Littleton, CO

[13]

Smart Homes, Smarter Consumers – Eric Elias

THE SMART HOME MOVEMENT is taking the world by storm. It's already huge when discussing home electricity usage, but smart home technology has not yet gained widespread acceptance when it comes to water. This is due to a number of factors that have literally left water underground from the public's point of view: water usage is only valued or noticed when a homeowner has a leak, gets an unexpected large bill, or when water is turned off for maintenance.

Homeowners didn't understand their water use or what is going on within their homes – that's changing, now that water rates are rising. And this issue has three parts to it:

- rising rates will force consumers to care more about their water use and how it will affect their wallets
- assets are exposed through the possibility of leaks and flood damage, and
- specifically in areas where drought or population are stressing available resources, conservation and awareness will start to become part of the norm.

Rising rates will encourage the adoption of smart systems

Circle of Blue found, through a survey of water rates in 30 U.S. municipalities, that rates have increased an average 33 percent since 2010[1] – that's an enormous amount of increase on a percentage basis. As these rates rise, the economic incentives to understand water bills are coming to the forefront. The EPA estimates that 70% of water use occurs in the home, mainly from the toilet, clothes washing, shower, faucet, leaks, and other uses.[2]

Many water bills arrive with the total water use and a dollar amount associated with the use, but this is like getting a credit card bill without an itemized list of what you spent and where you used your card. Unfortunately, many utilities do not currently possess the technology to tell you how you are using your water.

Smart meters and systems understand the flow profile of different water uses and can help homeowners learn how much water they typically use. Similar to the smart technology for saving electricity, these systems will get wiser about water usage and homeowners won't be surprised by the next water bill.

---

[1] http://www.circleofblue.org/waternews/2014/world/price-water-2014-6-percent-30-major-u-s-cities-33-percent-rise-since-2010/

[2] http://www.epa.gov/watersense/our_water/water_use_today.html

LEAKS CAN BE REDUCED OR ELIMINATED WITH SMART TECHNOLOGY

Increasing rates can impact your wallet, but leaks can ruin it in a completely different way because slow leaks in homes can cause a drastic increase in water bills.

The EPA estimates that 10% of homes have leaks above 90 gallons[3] – how do you put 90 gallons a day into perspective? It's a fair question when many people don't usually know the amount of water they use per day. However, when the U.S. Geological Survey estimates that Americans typically use between 80 and 100 gallons per day,[4] it means that 10% of homes are using enough water for an extra household member.

This is as if someone would go to the store and purchase extra servings of food, but only to throw them out every day – it's ludicrous that this happens, but smart technology could help prevent it.

However, larger leaks pose a different issue and can cause major damage if they aren't caught early. A basement flood causes both financial and sentimental damage, and these leaks can destroy anything on or close to the ground, even causing damage to large sections of the house or basement itself.

This occurs whether the leak is in a basement that is used regularly or in a vacation home that may only be used seasonally – large leaks can happen overnight or during the off-season months. Luckily, leaks that cause damage are now able to be caught by smart metering technology, which can notify homeowners immediately so that they can take action to prevent damage.

3
    http://www.epa.gov/WaterSense/pubs/fixleak.html

4
    http://water.usgs.gov/edu/qa-home-percapita.html

CONSERVATION AND AWARENESS ARE ON THE RISE

Conservation is a hot topic these days in parts of the U.S., especially in Texas and California. How can we affect behavior change to reduce daily water consumption? By showing people their water use.

In a book titled *The Smart Grid for Water*, Trevor Hill and Graham Symmonds note that automatic meter infrastructure (AMI) has contributed to between 9% and 15% reductions in water use, with 14% to 34% as the extreme example in Enid, OK.[5] How can we build on this to make it "fun" for consumers?

There are a lot of companies on the leading edge of gamifying behavior. Wearable technology, such as the Jawbone Up or Fitbit Flex, allow people to track their movements and sleep habits in a fun manner. Smart water networks can do the same, which empowers households to be smart about their water use. Customers can compare (or compete) with demographically-similar groups in order to affect behavior change, and they can learn about water-conserving tactics or water-efficient products, such as low-flow toilets.

SO HOW CAN WE RESOLVE THESE ISSUES WITH SMART TECHNOLOGY?

There are many companies working on a solution to help consumers solve problems with rising rates, leaks and conservation methods. Some, like Fathom and WaterSmart, are tapping into the growing use of advanced meters being deployed at the municipal level.

---

[5] Page 128. Hill, T and Symmonds, G. The Smart Grid for Water. 2013. Advantage Media Group. Charleston, South Carolina.

But what do you do if your municipality has not partnered with one of these companies or your meter is not part of a smart system? There are now several options for consumers, such as:

- a product called Leak Defense that monitors the water main and alerts ratepayers about flows that are consistent with leaks,
- Sprav is an interesting sensor that attaches to a shower-head and sends alerts to a homeowner's smartphone the longer the shower runs, and
- Lagoon has developed a sensor that wraps around the main pipe (to learn a household's typical usage) and it allows consumers to learn about their water behaviors and provide behavioral incentives for conservation – the smartphone app will even alert a homeowner if it detects a leak.

It's a great time to be involved with smart water technology because a better understanding of household water usage is starting to change the game: as rates rise, people want to see the details of their water use and learn about their behavior. At the same time, leaks can cost money or damage homes, so smart technology can help homeowners reduce the risk of a major incident. And, with conservation goals on the rise, people living in drought-ridden areas are looking for ways to change their behavior; with smart technology in the home, this can be accomplished with ease.

For utilities, homeowners and water entrepreneurs, the household usage of water is too important for us to continue to fly blind. Using smart technology and smart water systems will help us preserve our precious resources and encourage consumers to take a more proactive approach with water itself – after all, it's the smart thing to do.

About Eric

Eric Elias has been working in product development, business development and finance for more than 10 years. Prior to becoming CEO at Lagoon, he has worked at a variety of companies, such as SoMoLend, The Nielsen Company and General Electric. Eric earned his bachelor's degree in Finance and International Business from Washington University.

[14]

Why We Need to Unlock the Enormous Power of Data in the Water Industry - Sunit Mohindroo & Ahmed Badruddin

AT THE HEIGHT OF THE WORST DROUGHT in her generation, a young college student was looking for answers. Erin was intrigued by all the media attention focused on the drought and the water concerns that surrounded her college in Southern California. She wanted a better understanding of the local water supply and what shape the water system was in.

So she went digging... but she didn't know where to begin. Erin simply ran an Internet search for "water data" and found a handful of links, but these databases were fragmented, out of date and didn't seem very credible. Even reaching out to local experts and state agencies resulted in little information because no one really knew where it was or if the data actually existed; if it was available, who knows if anyone knew how to access it.

For many of us, there are only a handful of simple, straightforward questions, such as "what's the status of the water resources in my area" or "what types of treatment were in place to ensure water quality". After days of online searches, phone calls with local water authorities and a few emails, Erin was forced to accept the fuzzy, not-quite-complete picture of her water system that the news media was portraying.

How is it possible that we live in a time when you can surf the Internet and watch live television in the palm of your hand, but we haven't implemented better ways to produce, analyze and deliver useful data about our water systems?

This information void started it all

A few years ago, we were also curious (like Erin) about our local resources, so we started talking to people in the water industry - from farmers to treatment plant operators to scientists - to see what types of patterns would emerge and what the problems were. After our conversation blitz was over, we talked to more than 500 industry leaders from all over the world.

What we found was more than just a set of data challenges – we found a void of actionable data within the industry. That provided the impetus for WatrHub Inc., giving two software geeks the chance to pair up on a project that would impact society, primarily through bringing actionable data to the forefront of the water industry. We wanted to speed up innovation adoption and eliminate the lack of transparency within the world of water; now is the time to unlock the true power of our sector's data.

CURRENT STATE OF DATA TODAY

Before we can solve data challenges in the water industry, we first have to look at the current state of data within the industry today.

There is a lot of water data available, but much of it are mandatory documents for regulatory agencies. Seasoned water industry professionals would tell you there isn't a lack of data out there, although they would point out that there is a void of timely and actionable data that could really have a revolutionary effect on water's future.

PAPERWORK IS STILL IS IN THE MAJORITY

Most of that mandated documentation is connected to the U.S. Clean Water Act, which dictates a significant amount of the regulations for water and wastewater. There have certainly been changes to how we regulate water since then, but many of the reporting processes have remained with the same mix of paperwork and archival work.

Paper forms are printed out by water plant operators, filled out by hand and sent in to a local office for verification. If those documents are in satisfactory order, then they get shuffled together and are archived with documents from other plants. Most of us can understand the idea of having a paper copy in storage, but the information is worthless if it's not easily accessible to journalists or researchers.

Imagine if renewing your driver's licence had a similar process: two weeks to gather your driving record documents and another two weeks to manually review those documents. You could only imagine the frustration for a basic licence, so think how that could relate to an entire regulatory system.

Now, we don't want to paint a picture that is too draconian because there are state water agencies and watersheds around the continent that have converted to digital repositories and filing systems for regulatory documents. This is an excellent step in the right direction, but this process is primarily the digitization of hard copies, not utilizing this technology to create a searchable, accessible (even real-time) data platform.

Data is fractured and unstructured

Beyond accessibility issues with the paper format, today's water data is fragmented, incomplete and lacks a reasonable structure – missing information within these datasets can be very troublesome for any type of research.

Data scattered across hundreds of thousands of sites across the U.S. presents a major challenge for anyone wanting to make meaningful and timely observations on that data. You always want to work with a set that is statistically significant in size, but you also want to know that there aren't large swaths of data in a hundred other locations that could completely alter your conclusions.

Lack of standardization causes confusion

Every state in the U.S. issues its own driver's license with its own interpretation of what that plastic card should look like, but the positioning and formatting of personal details differs for every state. It's the classic dilemma when someone checks an out-of-state license to see if someone's of a legal age because the data is not standardized across each state.

Now, you can't really fault different water jurisdictions for presenting their mandated data in different ways, but the same type of dilemma occurs when someone's looking for a specific effluent limit in discharge permit documents from 10 different states.

Unstructured data can take many forms, such as water effluent limits written in paragraph form (instead of tabular form) or any numerical data that is not present in spreadsheet format. Similar issues create a stifling limitation to calculations and sorting functions that most analysts use. If the original data being analyzed is not standardized or structured, then sorting or filtering may have poor results (if any) because we would ever get the full picture from an unstructured dataset.

When Superstorm Sandy hit the Eastern seaboard in October 2012, state officials were scrambling to find out the conditions of water and wastewater treatment plants that were damaged in the storm; this was to ensure that the systems were operating within their normal parameters. Unfortunately, there wasn't a structured or easily-accessible database of operational data to get the plants up and running within a matter of hours instead of days or weeks.

Water innovation is also affected by the lack of standardized information, especially for companies that are looking for ideal candidates to pilot newer, more efficient technologies. Instead of cold calling every plant operator, someone can use a centralized information hub to pinpoint the facilities that would immediately benefit from a specific product; we hear about this type of scenario quite often in the water industry.

Think about data that isn't available or too outdated to be useful...would Google or General Electric base their current market strategies on very limited data or "current" information from five years ago? Not at all, but it's amazing we expect our water entrepreneurs and technology vendors to accept this.

In the end, all these factors contribute to one major challenge in keeping our water secure and sustainable: inaction by our policy makers, operational leaders and water entrepreneurs. And the lack of useful data, which can help them make decisions quickly and effectively, is the reason behind this inaction.

The Future of Our Sector with Data at the Forefront

Real-time predictive information can become available to everyone, similar to weather forecasting. This would proactively solve our water treatment problems, such as sewer overflows, contamination events and equipment breakdown, before they become public concerns.

This is just one of the many ways that unlocking the power of actionable data can help revolutionize and modernize the water industry. On a grand scale, if that predictive information can help us forecast a drought, then we can use our easily accessible water system data to redirect our water resources and prepare for weather crises better than ever.

Other useful ways that we can harness this power include:

1. **Water and wastewater discharge permits** – if this information was made more accessible, plant operators could look to their peers for advice on equipment upgrades and maintenance on similar technology, and engineers designing new treatment membranes could quickly connect with facilities in need.
2. **Annual reports and budgets** – this information can show trends in future spending, allowing water entrepreneurs to focus their efforts on meeting the water challenges of interested municipalities.
3. **Discharge monitoring reports (DMR) and compliance reports** – if accessible and complete, this will show the trend of a facility's water quality and will bring a new level of transparency to the local water system, allowing plant operators to proactively address any issues before penalties or regulatory bodies get involved.

4. **Benchmarking** – combining each report type that we mentioned above (permits, annual reports, monitoring reports) into sizable, structured sets creates the ability to seamlessly compare facilities.
5. **Accurate predictions** – if we take localized water system data and overlay it with other local datasets such as weather, population dynamics and industry movements over time, we can improve urban planning, disaster mitigation and overall economic development of a region.

An example of the type of insights that could be harvested from Annual Reports is presented on the next couple of pages with data from over 450 cities in California representing 12 years of spending, with highlights of a few interesting situations.

How Do We Start Unlocking The Power Of The Data?

Looking at the current state of water data and the potential the data holds, we have developed a system over a number of years with the help of experts to unlock the power of water data. The system uses complex software algorithms and proprietary data processes to accomplish our goals, but it can be distilled into a few important concepts that anyone can put to use.

One Size Fits None

This is key to harvesting raw data: if you haven't truly identified the needs of your data's audience, then you might embark on an analysis mission that isn't at all relevant.

118 | *Damned If We Don't!*

Visit http://www.wateranthology.com/DamnedBook/Resources
to view this image in detail

About the data

This visual shows the percentage of city budgets (combined CapEx & OpEx) that are allocated to each city's Water Utility Department for each year between 2001 and 2012. It identifies and highlights unique spending scenarios that might otherwise would be buried and lost in a large dataset. There are over 450 spending trend lines layered into this one visual, each representing a different city in California. Using our visual technology, we show which cities have made water spending a priority and can uncover sales opportunities for you going forward. This information was aggregated by WatrHub Inc. The cities mentioned were chosen as reference points and were not meant to be interpreted as authorized case studies.

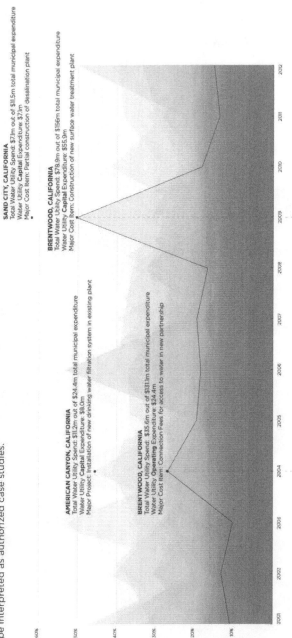

Ideas for Accelerating Change Around Water | 119

Profile: City of Brentwood

54 THOUSAND
2012 TOTAL POPULATION

$107 MILLION
AVG. TOT. MUN. BUDGET

$13 MILLION
AVG. OPERATING COSTS

$9 MILLION
AVG. CAPITAL COSTS

The City of Brentwood is a mid-sized municipality located in the East Bay Region of the San Francisco Bay Area. Over the past two decades, the City has seen two major spikes in the relative size of its water utility budget. In 2004 Brentwood needed to find an alternative source of water to meet growing needs. Previously the City received water owned by the Diablo Water District which also served the nearby City of Oakley. However, with Oakley's growing population, there was less capacity available to Brentwood. As a result, Brentwood required water supply and connection to the Contra Costa Water District, at a cost of $12,975,695. In 2009, the City completed construction of a new surface water treatment facility. This project was enabled by a major capital grant result, Brentwood required water supply and connection to the Contra Costa Water District, at a cost of $13.0. In 2009, the City saw its second water budget spike as it completed construction of a new surface water treatment facility. During the same time frame, the City approved and issued $53.0m in Water Revenue Bonds, which were in part responsible for the 349% increase in water utility revenues in 2009. This influx of funds helped enable the massive capital expenditure required to complete the new facility.

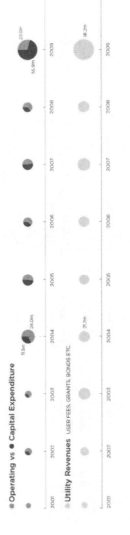

● Operating vs ● Capital Expenditure

● Utility Revenues USER FEES, GRANTS, BONDS ETC.

Sources: California State Controller's Office, City of Brentwood, City of American Canyon, City of Sand City

Data Mine with a Purpose

Use your audience's identified needs and criteria as a map for data hunting. Your mission is to explain the data that has relevance, not merely explore it; a lot less energy is wasted when you can zero in on a few critical characteristics and block out the rest of the data as noise.

Visualized Data is Actionable Data

How you present your findings to your audience will be just as critical as the data itself. Effective visualizations can range from filtered and prioritized spreadsheets to color-coded geospatial heat maps; choose the visualization that will allow the audience to easily understand the relevant data. This whole process will ultimately lead to better decisions based on actionable information.

So what happened to Erin and her search for water data?

Erin, the college student in Southern California, contacted us at WatrHub Inc. in the midst of her water data search after hearing about us on CNN. She was excited about our fresh take on water data and how we were approaching the data problems within the industry from a different angle.

We loved her enthusiasm for data; so much, in fact, that we hired her as a researcher on our data team at WatrHub. She is now helping us aggregate and deliver actionable information to the water sector because our industry needs all the help it can get.

Especially from the power of our data.

About the Authors

Sunit Mohindroo is the Chief Product Officer and Co-founder of Toronto-based WatrHub Inc. His strong computer background includes roles at Telus, Microsoft and Apple. Sunit has a bachelor's degree in Systems Design Engineering from the University of Waterloo.

Ahmed Badruddin is the CEO of WatrHub Inc., and he has worked as a software engineer at Microsoft. Prior to his role at WatrHub, he was a co-founder and the CEO of a smart grid startup company called Simpleafy. Ahmed has a bachelor's degree in Electrical and Computer Engineering from the University of Toronto.

[15]

Industrial Ecosystems: 21st century challenges, 21st century solutions – Jim Lauria

WE'RE AT THE CUSP OF A NEW INDUSTRIAL REVOLUTION – the era of the "industrial watershed" – and it is poised to change the water industry as well as industry in general.

For thousands of years, water supplies grew and consolidated, moving from communal wells to sprawling municipal water systems. We also built a massive infrastructure of reservoirs and canals, constructed huge hydropower plants, and developed vast networks of drinking water and sewer pipes.

But, in recent years, the benefits of decentralization are becoming clearer in many places. Sprawling cities – in both developed and developing nations – realize that they can augment their overstretched grids with decentralized micro-grids, which provides service to new neighborhoods with small, integrated water treatment plants.

And, as industrial facilities, from factories to factory farms, manage volumes of water – that could be as much as the flow in and out of full-sized cities – we have begun to recognize that they comprise their own ecosystems, complete with their own watersheds.

Industrial Watersheds Are Like Micro-ecosystems

Every drop within an ecosystem is accounted for, whether it's water that gets pumped in, flushed out or falls from the sky. By treating industrial watersheds as micro-ecosystems, we must make sure that the water is measured, analyzed, treated, used and released with the best possible outcomes in mind, including how each of those steps affects processes both upstream and downstream.

Industrial watersheds – as ecosystems within larger ecosystems – can be managed better think of them in the realm of organic design: the water system resembles the human circulatory system of pumps and pipes, the immune system disinfects the water, the kidneys filter it, a nervous system of sensors and (at the center of it all) a brain that coordinates all the functions to keep the system alive and aware of itself. Such biomimicry systems enhance the efficiency and resilience of industrial watershed management.

For instance, there are projects (by Geosyntec in Missouri and North Carolina) that use constructed wetlands and rain gardens to manage stormwater in urban developments. However, these biomimicry systems are equipped with brains and nervous systems. When weather forecasts predict rain, the Internet cloud comes alive with signals prompting retention ponds to drain, which make room to capture new water and prevent combined sewer overflows (CSOs).

Optimizing the water that flows in and out of industrial ecosystems challenges us to innovate. And when we operate sophisticated, cost-efficient industrial watersheds instead of massive geographic ones – managing one million gallons of water a day versus the amount for one million people – we can afford to innovate.

LOOK AT A WATERSHED AS A STRATEGIC RESOURCE

Military and political leaders have always recognized the strategic value of water: written in Sun Tzu's *The Art of War*, as noted by Hezekiah's Tunnel beneath Jerusalem's City of David, shown by battles over major river systems, and read about the current water struggles in Asia and the Middle East. Now, companies have started to take a strategic view of industrial watersheds as a valuable resource.

In a 2013 poll of industry leaders, the Pacific Institute and VOX Global found that 79 percent of companies reported that they faced current water challenges, and 86 percent of these companies expected to face similar challenges within five years. Water issues affected the bottom line of 57 percent of those companies, the same amount also said water would affect their growth within the next five years, and 80 percent of companies – within the next five years – expected water to impact wherever they will expand their operations.

Growth, supply and siting; if those aren't strategic issues, I don't know what is.

In any industry – whether it's food processing, beverage production, pharmaceuticals, mining, oil and gas (from extraction to refining), or power generation – water is integral to the business. Viewing any of those businesses as a watershed helps us reflect on water at every stage of its cycle within the process.

Water, as an ingredient, is used in finished goods. Service water transports products through the system – picture the flumes that float tomatoes through a cannery – or the water heats, cools and/or cleans the production system. Then there's process water that needs to be treated before it leaves the system.

Each of those gallons of water has a value and a cost; as we assess those values, we can determine the appropriate costs. Then we can fine tune the resources that we put into securing the water, treating it to the level necessary to do its job – cooling water doesn't need to be treated to the same level of purity as ingredient water – and preparing it for its journey downstream or, even better, back into the system for another round.

There's also the benefits and disadvantages of stormwater within the micro-ecosystem of our industrial watersheds. On the plus side, rain can be harvested as a free source of water that can be applied to tasks like cooling, flushing toilets or irrigating green space. Unfortunately, stormwater can also create huge challenges as it sweeps up contaminants and threatens to move them into receiving waters, or the flow can overwhelm wastewater systems with CSOs.

A scrap metal processor in the northwestern U.S. is a great example of the on-site management of stormwater in an industrial watershed. Stormwater, running through piles of scrap metal, was creating a pollution problem, so the company installed a treatment system. It reused the newly-treated water in cooling the company's crushing and grinding machinery, and it provided dust control; this process reduced the amount of contaminants in outflowing water and cut costs for fresh water purchased from the city.

The more efficiently we manage water within our industrial ecosystem, the more economical we can be. And the less water we draw or dispose of – each with its attendant costs – we reduce any water wasted. Now that's strategic.

21ST CENTURY CRITERIA

Growing populations and dwindling supplies has created significant challenges for water users in the 21st century, such as scarcity, pollution and politics.

Water scarcity is the central concern in many regions. It is creating unprecedented pressure for the wise use of water, inspiring new cost models that encourage conservation and spur the search for new sources of water. We can no longer take fresh, clean water for granted in many places, so we are beginning to create new supplies by treating wastewater, rainwater, brackish groundwater and even seawater.

But the roles water plays within each aspect of our industrial watersheds makes it clear that each micro-ecosystem might need different types of treatment. It's a good thing that existing technology can treat many water sources to a nearly-perfect level of purity, such as Singapore's remarkable NEWater toilet-to-tap system; the city's supply of tertiary-treated wastewater is cleaner and more reliable than other available sources.

Scarcity-prevention examples can be seen in smaller locations as well, such as the water treatment process in the city of Spanish Fork, Utah. By treating irrigation water to a lesser degree than drinking water, the city can use high-efficiency filtration units in buildings that are designed to blend into housing subdivisions. Then, the treated water feeds into pressurized irrigation systems to keep Spanish Fork beautiful and green, and it doesn't deplete scarcer or more thoroughly treated drinking water resources.

For pollution concerns, industrial watershed managers are beginning to clearly recognize the impacts and externalities of the used water they return to the system. Christopher Gasson, publisher of *Global Water Intelligence*, has pointed out that we have traditionally taken water from nature for free and returned it without concern for the impairments we cause.

The growing calculation of water footprints – a process that got increasingly complicated as ISO models began recognizing the vast array of impacts and links among externalities in water use – highlight the pollution and societal costs we add to every gallon at the end of the pipe. The very fact that we can't tie a discrete number to a facility's water footprint, as Gasson points out, is a strong indicator that our footprints are large and heavy.

Real-world management requires us to think in terms of footprints: it's not just how much water goes in or out of the system, but how impaired the water is at the end. It also forces us to consider the size of the water treatment facility's physical footprint, such as square footage or the amount of building materials. We also must think about the facility's chemical footprint, carbon footprint and energy footprints.

Efficiency will have to take into account all those measures. In many cases, it will become apparent that the most useful things we can do to reduce all those footprints are to use less water from the outset, reuse (or recycle) more water, and efficiently treat whatever leaves our industrial watershed.

AT THE END OF THE PIPELINE

Ultimately, most geographic watersheds flow to the sea, and most industrial watersheds have outfall, too. Regulatory structures in California – including Title 22 standards for releases of wastewater into the environment and total maximum daily loads (TMDLs) that account for all contributors to pollution – provide a strong indicator of what is in store for most water users in the future; these regulations certainly affect North American supplies and will likely produce similar legislation around the world.

Winemakers in many areas (such as California) are increasingly required to treat effluent before discharge into municipal sewer systems because low-pH, high-solids content can interfere dramatically with biological reactions at wastewater treatment plants. Fats, oil and grease are widely recognized for their destructive impacts on sewer infrastructure, and ecologists are much more aware of the impacts of nutrients and the biological oxygen demand (BOD) on receiving waters.

Ironically, even with all of these regulations to reduce California's wastewater conditions, Silicon Valley remains quiet on the subject, even though the high-tech world uses massive amounts of water –

think about cooling systems for servers, the construction of computer chips and wire, and the water-intensive process of producing energy.

That's why it's extremely helpful to look beyond the "borders" of the industrial watershed when grappling with wastewater issues. In Israel, where tertiary-treated wastewater provides half of the nation's irrigation supply, high levels of boron from household detergents – and not removed by treatment – were damaging crops when the water was re-used for irrigation. The Israel Water Authority sat down with detergent makers, and, shortly after those meetings, only boron-free formulations were available on the market.

Interestingly, as Israel introduces more desalinated seawater into its municipal potable water system, the quality of recycled wastewater is expected to improve. The local groundwater – apportioned first for municipal and industrial uses, then treated for use in irrigation – has an electrical conductivity of 2.0 to 2.5 microSiemens per centimeter by the time it's been used, flushed and treated. When desalinated seawater is used from the start, its low salt content will drop the conductivity of recycled irrigation water by half.

Recycled water has proven itself again and again, from Israel's agricultural powerhouse to the Beijing Olympic Village. Ultimately, toilet-to-tap projects, like the 70-to-100 MGD system in California's Orange County, will be emulated worldwide.

The Orange County project demonstrates the remarkable role of technology in purifying water that leaves the municipal (or industrial) watershed. Microfiltration is followed by reverse osmosis and backed up through ultraviolet (UV) purification. The tertiary-treated water is then leached into the ground, using the earth as a massive sand filter to ensure purity by the time supplies reach the local aquifer, which rebuild local water reserves and fight saltwater intrusion from the edge of the Pacific.

After centuries of reliance on old technology – with filters that date back to the Roman era and sand media systems that reflect the best thinking of the early 19th century – the water industry has now become a high-tech endeavor and we need to look at modern-day solutions.

Who will claim to be the Silicon Valley of water technology?

With 6 billion consumers, there's no greater global marketplace than water technology, but who will claim to be the leader? Israel has a huge array of leaders in the measurement, conveyance and treatment sub-sectors of the water industry, but Milwaukee, Pittsburgh, Colorado and Houston are also leaders in various areas of our sector.

There is no greater challenge for innovators of the water sector; it's not only a profitable venture, but an honorable one. Water is where technology meets life itself – most of us are linked to non-renewable sources, whether we're drinking and bathing in them or using the water to grow our crops.

Firms from Israel to Silicon Valley must begin to view water as their issue, their priority, and their cause. The enthusiastic and creative input of smart technology companies could propel the water industry to new heights and raise the tide for all boats. While the world's best high-tech minds build the nerves and brains of the next generation of water systems, leaders of industry can adopt the industrial watershed mindset.

A commitment to viewing industrial water holistically will yield better solutions. A commitment to considering water from the very start of an industrial project – from the siting to the design and the industrial engineering of it all – will yield more effective and more efficient systems. And a commitment to viewing the industrial watershed in the context of the geographic watersheds surrounding it will remind us why we need to take such dramatic steps to make every drop count.

About Jim Lauria

Jim Lauria is a water technology executive with a Bachelor of Chemical Engineering degree from Manhattan College. He has spent over twenty years selling process solutions to the brewing, wine, food & beverage, mining, oil & gas, power and other global industries. Jim has provided peer review for the World Health Organization's publication on drinking water treatment and his blog on the Huffington Post about worldwide water management practices has generated responses from all levels of industry and government. He can be contacted at jim@jimlauria.com.

[16]

Convergence: Utilities need to mind the data as well as the drops – Graham Symmonds

THE WATER CYCLE WILL BE THE MEDIUM through which we will first experience the impacts of a changing world climate. Increasing hydrologic volatility means increasing risk, dramatic water supply interruption events, increased financial pressures, and a significant chance of an imminent failure of our supply-based infrastructure.

While there is no single "silver bullet" to solve our water scarcity and volatility issues, utilities can begin to implement practices and policies that demand reduction through better, more informative rate structures and data; increase water reuse through the adoption of a "right water for the right use" concept; and improve agricultural and municipal water efficiency. This combination of infrastructure, information and incentives will be the hallmark of the smart water utility in the future.

Underpinning this drive to sustainability is a requirement to get more from our existing systems – do more with less so we can maximize our efficiencies. As with transformative change in other industries, this goal will be achieved through maximizing the use of data across the entire utility ecosystem. This convergence of data will not only allow for our utilities to operate more effectively, but will be a key driver in maintaining our infrastructure, meeting our current and future demands, and assuring the revenue stability of our utilities.

Utility Data Systems

Utilities employ a wide array of data systems in their operations. These can be broadly categorized as follows:

1. Customer Information Systems (CIS)
2. Customer Engagement Platforms
3. Meter Data Management Systems (MDM)
4. Asset Management Systems (AMS)
5. Geographic Information Systems (GIS)
6. Supervisory Control and Data Acquisition Systems (SCADA)
7. Hydraulic Modelling Systems
8. Laboratory Information Management Systems (LIMS)

These systems have grown organically, much like utilities themselves – typically from specific needs. As a result, many of these data systems are departmentalized, or siloed within the utility organization.

The engineering department needed to understand the operation of the distribution system, so that team invested in a hydraulic modelling platform. The operations department needed to track maintenance activities and brought in an asset management system. The customer service organization needed customer-specific information and implemented a customer information system. While these platforms operate to meet the needs of their masters, often there is little or no integration/data transfer between the utility data systems.

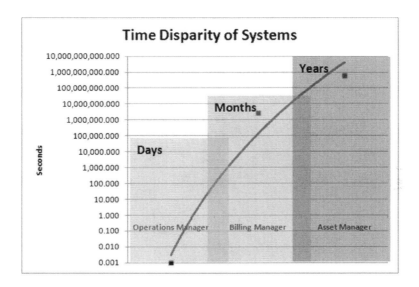

Complicating the matter is that each of these systems has a time frequency that may not be compatible with other platforms. This time disparity means that homogenizing this data to distill meaningful information is a challenge, especially when there are varying intervals: SCADA systems operating at millisecond intervals, customer information systems and billing platforms operating on a monthly cycle and engineering systems operating at decadal frequencies.

THE NETWORK EFFECT

The drive, however, is to increasingly more granularity in our data systems. And this is for a good reason because linking data systems provides significantly more information for the utility than does operating these systems in separate silos.

As we improve the connectivity between platforms, the information increases as a function of the square of the number of interconnections. If we consider adding connectivity to an existing data platform that increases the available nodes of data from 10 to 15, the information increases by a factor of 233%. This, of course, is a double-edged sword: increasing interconnectedness yields more information, but increasing information demands more powerful computation and data analysis requirements.

Fortunately, significant reductions in the cost of data storage and analysis resources available with the widespread adoption of cloud-based computational power and networking infrastructure are increasing the opportunity to collect, retain, and analyze large volumes of data at increasing data velocities.

The Value of Interconnected Data

The value inherent in interconnected data is manifold. The insights that can be derived from analyzing data from multiple data sources directly benefit the resource and financial health of our utilities.

Some basic ways that utilities can take advantage of these data streams include:

1. Using data to ensure the utility is billing for all water delivered

Much of the "leaking revenue" in utilities can be traced to incomplete or corrupt data systems.

Once installed, it is very common for the physical infrastructure – a utility's assets in the ground – to diverge immediately and continually from the logical infrastructure, which are those assets in the utility's data systems. Forcing and maintaining the convergence of this data is critical to the well-being of our utilities.

The problem for our utilities has been maintaining the alignment of data and data systems. Historically, it has been manually maintained by checking the infrastructure in the field, then taking the time to amend or repair the data faults. Unfortunately, those field results would get stale very quickly in the utility world, which would likely create more deviation in the data stream.

However, by combining data layers and elements from other sources, the majority of this activity can be automated to provide for validation of infrastructure in an on-going manner.

2. USING DATA TO MAXIMIZE THE REMAINING LIFE OF OUR INFRASTRUCTURE

Our water and wastewater infrastructure lacks sufficient investment to maintain it *in perpetuity*. However, combining data from various sources allows utilities to reduce the effects of leakage and reduce the pressure operating envelope in their pipelines.

Pressure management has become an established leak control methodology for water systems by reducing losses, but it can also be critical in extending the life of pipelines by reducing the occurrence of bursts. By reducing leakage, the volume of water necessary to support the utility decreases, which subsequently reduces the operating pressure required to maintain that flow.

A mere five percent reduction in flow reduces the pressure requirements by 9.75 percent – that has a dramatic impact on the amount of physical leakage and the potential for transient pressure peaks to generate main bursts.

3. USING DATA TO INFORM CUSTOMERS AND ENCOURAGE CONSERVATION

With integrated data – combining the elements of where, when and how water is used – the customer can begin to make truly conscious decisions on water use. Combining data from MDM and CIS, and by employing Customer Engagement tools, customers can get a real-time understanding of their water use.

This knowledge has an impact on the water resource as well as the costs they are incurring. By making more customers aware of their usage, the data can result in per capita demand reductions from 5 to 15 percent.

4. USING DATA TO ASSURE THE QUALITY OF WATER IN OUR SYSTEMS

Connected data systems can also be used to significantly improve the utility's understanding of water quality in the distribution system, and this can dramatically improve response times. By combining CIS, highly granular consumption data from MDM, operational information from SCADA, hydraulic modeling data and geo-referenced spatial asset management data, a rapid, graphical assessment of water distribution system health can be achieved.

5. USING DATA TO "FIND CAPACITY" IN OUR EXISTING INFRASTRUCTURE.

Integrated data systems are instrumental in finding capacity that is existing in our current infrastructure. This "found" capacity can result in significant deferral, even elimination of near- and long-term capital requirements, which can be used to realize significant reductions in operational expenses.

Conservation takes advantage of a utility's existing infrastructure, but – due to outdated regulatory demand figures or well-known rules of thumb – this capacity is usually sequestered from use. But, through MDM, SCADA, hydraulic modelling and CIS, a utility can liberate this capacity from "hibernation" by using real, location-specific demand numbers.

Additional opportunities for data convergence are shown in the table below:

CIS	MDM	LIMS	SCADA	Hydraulic Model	GIS	AMS	Customer Engagement	Use
*	*				*			Revenue Assurance
	*		*	*				Real-time Demand Projection
	*		*	*				Pressure Control
*	*	*	*	*	*	*	*	Proactive Water Quality
	*		*	*				Found Capacity
*	*						*	Customer Alerts
*	*						*	Demand Reduction
*	*						*	Conservation
*	*							Real-time Billing
	*		*					Virtual DMA
	*		*	*	*			Leak Detect
*	*		*	*	*			NRW Reduction
	*							Meter Health Monitoring
*	*		*	*	*	*	*	Distribution System Efficiency
*			*		*		*	Outage Response
			*		*	*		Maintenance Efficiency
	*	*	*	*				Fate of Contaminant Modelling
*	*						*	Ordinance Enforcement

Conclusion

In general, we do not suffer from a scarcity of data, but we struggle to turn that data into useful and actionable information.

However, by converging the disparate datasets currently in use, water utilities can significantly improve the understanding of the utility landscape. Advances in technological sophistication, open architecture, cloud-based systems and infrastructure, and the availability of more granular data offer huge benefits to our industry.

These structural, technological and physical improvements can drive significant efficiencies for utilities, which improve the operations of infrastructure, reduce customer demand, assure revenue protection and maintain a highly reliable, available and safe water supply.

About Graham

Graham Symmonds is the Chief Knowledge Officer of FATHOM™ and joined Global Water in 2003 as the Senior Vice President for Regulatory and Compliance. He has years of engineering experience utility oversight in corporate roles. He was also commissioned as an officer in the Canadian Navy. Graham is the co-author of *The Smart Grid for Water: How data will save our water and your utility* and has written several pieces on effective infrastructure design and usage in the water industry. Graham has a bachelor's degree in Mechanical Engineering from the University of Toronto and concluded his post-graduate education at the Royal Naval Engineering College in Plymouth, England.

[17]

The Hitchhiker's Guide to Digital Efficiency – Alan Hinchman

THE AGE OF THE INDUSTRIAL INTERNET IS UPON US, which means water and wastewater utilities are now taking advantage of newer technologies that are changing the landscape of our sector. With a few pieces of software, utilities can now optimize their operational performance, enhance asset reliability and accurately monitor their systems.

Some of the brightest minds on the planet are developing software for the next-generation of Industrial Internet technologies, and the potential advantages for the industry are nearly unlimited. These products allow for more intelligent management of assets by harvesting and processing massive amounts of data to produce actionable information.

The rate and complexity of this technological development can be both exciting and unnerving, even it may be unclear how everything fits together or what to do with all the available information. So how does one even begin to incorporate it all? Adapting to the technology can be tricky, but the first step is mentally preparing for this fundamental shift in your operational philosophy.

Getting started can be the hardest part, but it's also the most important step in protecting your operation from unplanned downtime that can directly impact the bottom line; when optimized, these technological changes could save your organization millions of dollars. It's better to start with smaller changes, especially for your employees – this allows them to gradually get comfortable with new processes before adopting more advanced technologies. Most importantly, a well-formulated plan will help your organization transform from a reactive and unpredictable operator into an efficient, intelligent and highly-profitable one where data is now your best business partner.

Monitor Your Assets

To kick off your journey, let's begin with basic monitoring of your most critical assets. This is done by equipping critical machinery with sensors and control networks that read and manage the actions of that specific piece of equipment. By collecting and simplifying the data produced, an operation starts to form a solid foundation that determines how all the separate parts of the system are supposed to function as a complete unit.

While experienced personnel might believe they are very good at catching and solving problems before they become much worse, data is much more accurate and more reliable – it never takes a vacation or sick day. In other words, data is the key to all the dirty little secrets your machines keep from you.

As the industrial process unfolds, the installed sensors pick up information about the speed of each step, the volume of liquids, the internal temperature of a pipe, and where a leak may be letting air or water into the system. The sensors and controls are doing the heavy lifting for you, which eliminates any doubt about what is happening within the machines. Once the data is harvested and funneled to one location, engineers can now watch and focus primarily on the data from an entire operation; they can access the information in a central office or remotely.

One way to access the system data is through an online Human Machine Interface/ Supervisory Control and Data Acquisition (HMI/SCADA) system at the point of control. This technology allows operators to view all systems in the process from one location, giving them greater access to real-time information and allowing them to make more informed decisions.

The other way to evaluate the data is through the cloud, which allows access from an established control center, but it also enables engineers to examine real-time data from a wireless Internet connection; this could be within the control center or out in the field. Now, any data being collected on the sensors can be analyzed at a central location, and controllers can remotely manage their assets.

Mobilize Your Workforce

Let's take monitoring a step, or many miles, further. Using mobile applications, engineers have front line access at their fingertips from anywhere they can connect to the web. Engineers can log in using apps on their iPad, pull up data for all systems under their watch, and monitor real-time information. Insightful and actionable information is always at the ready, so informed decisions can be made without delay, which helps avoid disastrous results and reduce unexpected issues.

Regular maintenance and upkeep will still be required, but the technology definitely helps keep small surprises from becoming major disruptions. Think about the cost avoidance from preventing a water line break: those costs alone makes it well worth it to know ahead of time when things might be starting to break down.

Fixing problems based on data and generally catching problems before they become catastrophic is beneficial on its own, but it still requires constant attention; even manual control leaves a lot of room for error. If that worries you, then you're ready for some advanced analytics to take you to the next level of protection.

Diagnose Your Pain Points

When analytics come into play, the benefits of intelligent data come into focus – this is when your organization begins to learn the answer to *"why?"*

By deploying an effective analytics strategy, you can begin to understand the data at a deeper level. Even if history wasn't your favorite school subject, you'll learn to love it now because your data history holds the key to preventing major disruptions and optimizing your systems in the future. Instead of physically mobilizing the maintenance team, trained personnel can check the history of the data to discover the root of the problem. Once identified, the team will be able to write a new analytic to give warnings next time a similar data trend begins.

Predictive software works in a few ways. Some systems use cluster analysis or similarity-based modeling. Others use Proportional Integral Derivative (PID) loops to save millions of dollars in unneeded equipment wear-and-tear that can result from large-scale process variability. These systems identify normal equipment behavior as they predict future behavior. This level of intelligence takes numerous variables into account, so these highly-complex relationships (with the data that accurately predicts future states) are based on any particular variation in one or more inputs.

Here's how it all happens. Imagine you're the only person who drives a given road every day to work – there's no traffic, no lights, no turns. This is how a machine *would* work if no other factors came into play, but it's never that simple. As the area around your commute gets developed, you start to encounter stop signs, crosswalks, construction that blocks a lane or two, and (usually when you're running late) you get stuck behind a garbage truck. Even though every day your drive is different than the last, you gradually learn from experience, so you'll know how to react to each future interruption for a safer commute.

This is essentially what happens with advanced analytic modeling. New input variation – new equipment, new materials, new processes, temperature and seasonal changes, and so on – has an effect on the function of each dependent system. Engineers can trust that a certain fluctuation of one input will have a standard reaction in a system down the line.

For example, a drilling operation in a harsh environment with volatile weather patterns would have an established analytic model that determines what the proper internal temperature of a machine should be, compared to the outside temperature. This allows for variation in "proper" internal temperature and makes it easy to distinguish when any deviation is acceptable or if it requires further examination. The frozen winter months in Michigan will keep internal temperatures on the low side, but it's a different scenario when everything starts to thaw: you will know how strong the correlating rise in machine temperature is supposed to be, or if the deviation is too extreme and signifies a real threat to the system's stability.

Learn from History

They say those who ignore history are doomed to repeat it, so why limit how much history you can reference? Oil and gas production produces a massive amount of data that is eternally useful; limiting your amount of data only limits your ability to make informed decisions. Industrial Big Data gives you the freedom to mobilize much more information than previously possible.

With the right applications, you will be able to gather data from more sources over a greater duration of time than ever before. You can compare all this data to your real-time activities to make quicker, more informed decisions than ever imagined. In addition to the emerging technology, cloud-based applications are a major factor with Industrial Big Data because massively scalable storage environments can collect from (and disperse to) variable locations give you the flexibility to work on more data and do it faster.

For example, solutions built on Hadoop®, can handle large data sets by clustering large numbers of low-cost commodity computers together to act as a single storage entity. Historian software allows for economical storage while maintaining rapid access speeds. These solutions scale vertically, for any velocity, as well as horizontally, for any volume and variety.

Optimize Your Operation

Now that your operation is working smarter and leaner, and you've built your trust in predictive analytics, you'll be ready to move further down the efficiency continuum. More sophisticated software analytics won't only catch problems before they happen, but they also allow you to standardize your process across all parts of the operation to optimize performance and value. Electronic instructions are provided to operators to ensure that the best methods of work are followed every time at every place, regardless of experience or training.

The most modern predictive analytics software available now allows you to move from self-defined analytics to a stage where machines learn on their own and constantly write their own analytic rules. You could call it artificial intelligence, but it's a very realistic shift from unplanned to planned downtime, which could potentially save your company millions of dollars. Predictive-analytic software reduces maintenance costs and improves asset lifetimes to a greater degree than human monitoring could ever do.

Yes, we will still require human interaction and maintenance, but some things simply can't be done manually. Advanced software can adapt to variations much faster than humans, where the system sends signals to automated controllers that keep everything running in optimal balance. The point of control goes from the HMI/SCADA system (or iPad app), and becomes starts to become embedded into the controls hardware itself. The information runs in a closed-loop system that greatly reduces the human element, requiring manual interference only when a major issue arises.

Now Dive In!

When all layers of software are incorporated into your maintenance and oversight strategy, you have reached your destination: you are a lean, mean, data-using machine. By taking each step one at a time and steadily marching towards the goal of automation, your oil and gas facility will be at peak performance in no time. Without having to worry about managing emergencies or wasting money on unnecessary procedures, think how much better your job will be at a safer, efficient and more cost-effective operation.

About Alan

Alan Hinchman is Global Water Industry Manager for GE Intelligent Platforms. The municipal water industry has always been a passion for Alan, and this role has allowed him to spend time helping utilities of all sizes around the world provide better services for their customers. As an expert is water distribution information technologies, Alan has been a frequent speaker at many industry events.

[18]

Non-Revenue Water: A prime target of water innovation – David McGimpsey

ADVANCED METERING INFRASTRUCTURE (AMI), big data and other elements of the smart water grid shine a spotlight on a number of issues – infrastructure problems, incorporating data into capital planning, early leak detection and so on – but many efforts are focused around a core issue for water utilities: non-revenue water.

A deep dive into non-revenue water metrics results in a treasure trove of useful information for utilities. Some would argue that these details provides greater value than any other metric because non-revenue water provides a window into the utility's management performance. Typically, the higher the non-revenue water percentage, the more likely it is that the utility engages in some or all of the following behaviors:

- significant deferred maintenance,
- sub-optimal funding of its main and meter replacement programs,

- failing to dedicate resources to leak detection, and
- failing to aggressively defend against water theft.

Few, if any, water utilities report non-revenue water figures that approach zero – overspending to reduce non-revenue water simply is not a problem. However, the discussion cannot simply focus on spending enough to achieve an efficient level of non-revenue water. A high non-revenue water percentage likely – but, theoretically, not always – indicates potential management problems, and water utility managers need to collaborate on ways to achieve optimal levels of non-revenue water in order to keep utilities running efficiently and limit the risks of deferred maintenance.

Some History on Non-Revenue Water

In the late 1990s, the International Water Association (IWA) formulated the parameters of non-revenue water to establish a standard nomenclature and methodology around non-revenue water. Calculations for water loss had previously consisted of guesswork and no homogeneous standard existed to allow for benchmarking.

Since the IWA announced the non-revenue water standard, known as a "water balance," many national water associations around the globe have adopted the non-revenue water standard, including the American Water Works Association.

Non-revenue water consists of three basic categories:
- Unbilled Authorized Consumption consists of water consumption that public policy exempts from paying rates, such as water used to fight fires or to flush mains;
- Apparent Losses (also known as commercial losses) accounts for water theft, data errors and metering inaccuracies, which

as exemplified by slow or misread meters and unauthorized usage of fire hydrants; and

- Real Losses represent the leaks in the system, from storage tanks to transmission mains, from distribution mains to pre-meter leaks.

Of course, non-revenue water will never be eliminated because mains will always need to be flushed, and – until we find an acceptable way to bill it – water used for fighting fires will be considered non-revenue water. Plus, complete elimination would be inefficient, especially if you think how expensive water service would be if the system was so tight that not a single drop ever spilled.

The important point is that utilities need to strive for an *efficient* level of non-revenue water, which can be achieved with the data acquired within our smart grid technology. And this process can help conserve water while reducing our costs, specifically those from non-revenue water.

Non-Revenue Water Metrics Save Us Money

It's no secret – America's water infrastructure is old.

And the longer we delay upgrades or significant maintenance to our systems, the greater the risk that utilities fall into a downward spiral of growing expenses. The more we defer maintenance, the deeper into disrepair the utility's infrastructure sinks and that ultimately costs ratepayers tremendous amounts of money to fix; the rising costs apply to repairs or replacements. The aging of water infrastructure naturally leads to main breaks and other leaks – significant sources of non-revenue water – as well as lost revenue from service disruptions.

This is why having a good main replacement program will reduce non-revenue water, avoid service disruptions and optimize infrastructure. Some utilities will upsize the pipe being replaced to handle new growth in the area served by the main, or other utilities will replace and upgrade pipes made from inferior or obsolete materials.

In addition to pipe replacements, water meters are a key factor in calculating non-revenue metrics because aging meters develop inaccuracies that favor customers (who may not be aware of the issue). This is why a meter replacement program aids in reducing non-revenue water by preventing significant quantities of water – that won't get registered by the meter – to pass through. It's important to note that meter replacement programs, especially for large meters, often have very short paybacks. Sometimes, this can be less than one year – that's a lot of water that the utility was giving away for free!

Another crucial way to limit non-revenue water expenses is by dedicating resources to detect leaks; if you look at the metrics after repairing major leaks, you will see how it pays dividends to the utility. By implementing a leak detection program, water utilities can find small leaks and monitor them before they grow into bigger and more significant leaks. Proactive leak detection dovetails with the utility's main replacement program by providing a valuable source of data on the distribution system, which then can be used to provide unique insights into what mains likely need to be replaced.

Finally, utilities that do not defend against water theft leave themselves susceptible to ever-increasing amounts of water being stolen. Develop programs to dissuade potential water thieves: identify illegal and unmetered connections, monitor temporary hydrant meters, and monitor any tampering with those hydrants.

While all of these steps will help reduce the costs associated with non-revenue water, the main reason for effectively using data in our operations is simple: to conserve a limited resource.

Non-Revenue Water Matters Now More Than Ever

Of all the water on Earth, only 3% is fresh water and 70% of it is under the polar ice caps. That leaves us with only 0.3% of this precious resource as fresh surface water available for human consumption... and future supply will become less scarce in the future. It's not that there will be less water on an absolute basis, but growing population numbers will reduce the average amount of available water per person. Climate change will also contribute to stress on water supplies.

With more utilities to deliver water to more people, there's also a larger chance that non-revenue water will increase. Several case studies and white papers discuss how water infrastructures in the United States, Canada and the UK are relatively adequate when compared to other countries; it's quite shocking to see how much non-revenue water (on a global basis) is being wasted.

As such, the quality of water available becomes a critical factor, whether the water is polluted or increased urban development causes greater watershed stress. Other concerns include: faster run-off, which increases erosion; sediment and nutrient deposition, which could affects filtration and distribution; and rising water temperatures from the inflow of warmer run-off water because this increases the likelihood of invasive species introduction as well as the rate of evaporation.

Non-revenue water could save utilities in the future because represents a tremendous source of supply that can be used, so utilities need to focus on this resource *now*. Non-revenue water doesn't need a well, won't need a reservoir and there's no need for additional treatment. The resource exists within the utility's control, but it escapes through categories of use identified in the water balance. The utility only needs to capture the greatest percentage of non-revenue water that is economically feasible.

Maximizing the Effect of Big Water Data

The power of AMI manifests itself in data use, which can be used in multifaceted ways, and reducing non-revenue water through AMI data analysis represents just one application.

Several effective uses of data from AMI include:
- developing sound rate design,
- automating meter reading and billing,
- avoiding service calls,
- orchestrating shut-offs or turn-ons remotely,
- and identifying leaks or other inefficient water usage behaviors.

Of course, these are all important, but this chapter is only focused on non-revenue water, its relationship to AMI and the innovative use of its data. One significant example of this innovative usage is by accurately detecting leaks throughout the entire system, not merely at main distribution points.

Leaks that occur before the meter are labeled as non-revenue water, and they represent real losses because that water is produced and fully paid for by the utility. However, with the data from AMI, the utility would be able to determine where the leaks occur throughout the system – this allows utilities to make repairs that would immediately affect the bottom line.

Moreover, if a utility's AMI data analysis identifies mains that have multiple leaks, then the information should feed into the utility's capital planning program so the troublesome main could be scheduled for replacement before a more serious rupture occurs. Scheduled replacements reduce non-revenue water costs because the utility will avoid losing water from a main break if the system deteriorated too long.

By taking a proactive approach to replacing a troublesome main, this action will ripple throughout the utility because it improves the utility's bottom line and increases customer satisfaction. Utilities will continue to sell water to customers that would be affected by a lengthy service outage, thereby boosting utility revenues. A temporary service outage for a replacement would be tolerated much more than an emergency service outage, so effective planning eliminates complaints and increases customer satisfaction.

Not only can AMI can be used to identify leaks in the distribution mains, storage facilities and transmission mains, but the AMI can also help locate where unbilled connections exist within a utility's system. Non-revenue water is not a value that utilities would like to admit, but many utilities have their share of connected (yet unbilled) consumption on their systems.

AMI Could Lead to Profits from Non-Revenue Water

For example, I was involved in a sale of utility assets, but there was a surprise to this transaction a few years later. The acquiring utility had identified a significant number of connected, yet unbilled, users on the system. The worst part? They had to guess where these water thieves were located because the acquiring utility didn't even have an AMI system!

Think about that asset sale and if the selling utility had an AMI system in the first place. Not only could they have identified and billed those users, but the utility could have collected more revenue and could have received greater total compensation in the sale. Unfortunately, because they did not know about those potential customers – who were not paying for water – the selling utility lost the opportunity to earn recurring revenue and earn the value of serving those customers when it sold the assets.

However, this non-revenue water situation had a silver lining for the buyer, especially due to the customers it had unexpectedly acquired; the data exceeded a reasonable amount of usage for the number of customers, which led to an investigation. Once the utility converted these apparent losses – thanks to non-revenue water data metrics – into billed accounts, the acquiring utility generated a new substantial source of revenue. Just imagine how much more usage could have been billed if the purchasing utility had an AMI system to identify these unbilled service connections much sooner... and who knows if this would have been profitable enough so the utility never went up for sale.

This is why AMI, along with the data that reduces non-revenue water, merely scratches the surface of how it can give a lot of power to a utility's future. This type of innovation in the water sector needs support to grow and develop, and those innovations will open up an entirely new supply for utilities: non-revenue water.

By using AMI and the powerful data it generates to reduce non-revenue water to efficient levels, the utility captures water that it already has produced and paid for. Capturing that "lost" water and being able to sell it to customers is much better for everyone; it's truly a waste when water slips through cracks and seams.

Preventing those losses are critical as we enter into the era of water scarcity. Now that's something everyone – from ratepayers to utility executives – can agree on.

About David

Dave McGimpsey practices utility and energy law with Lewis Roca Rothgerber in its Denver office. He has represented water utilities large and small and all sizes in-between since 1998. Dave also created The Water Values Podcast, available on iTunes, Stitcher and other podcast directories, and has hosted it since inception.

[19]

Tapping into the Power of Behavioral Science: Insights & Opportunities for Water-Use Efficiency – Jeff Lipton & Caroline Hodge

THE WATER SECTOR NOW EMPLOYS BEHAVIORAL SCIENCE to persistently nudge residential customers to use water more efficiently.

Many conservation and efficiency efforts have focused on providing financial incentives for consumers to reduce their resource use, but consumers have not invested in cost-effective, efficiency-driving products or practices as widely as expected.[1,2]

Providing consumers with information, which include steps they can take to reduce their resource use, also fail to produce measurable behavior change. Studies indicate that providing consumers with information alone can increase their knowledge of a topic, but rarely produces significant changes in behavior.

One study, for example, showed that individuals who participated in a workshop on residential energy conservation showed changes in attitudes and knowledge, but no changes in behavior.[3] In another study, individuals who took a 10-week course about conserving water showed no subsequent changes in water consumption.[4]

Over the past decade, utilities, governments, businesses, and non-profits have come to realize that more than just financial considerations and information drive behavior. Social and psychological factors also play a significant role in shaping consumers' decisions and behaviors around resource use. Utilities can use these factors to create a drive toward effective water management through the power of social norms.

Behavior will gravitate toward the perceived norm.

Leveraging social norms to change behavior

People often actively resist the idea of being influenced by social norms. Despite dramatic conformity in many experiments, participants often tell researchers that they came to their conclusions independently of the beliefs and behavior of those around them.[5][6][7] And a recent study found that consumers did not want a social comparison – a "show me how much [energy] I used compared with households like mine" type of statement – to be used as part of an energy efficiency program.[8]

So it's quite ironic to see how making people aware of norms, especially when these norms correct a misperception, can have an enormously powerful effect on their future behavior. Numerous studies have shown that making people aware of what is truly "normal" among their friends, peers, or neighbors can make them more likely to adjust their behavior to match that of those around them.

Social norms have successfully motivated behavior change in a range of other areas, such as smoking, voting, retirement savings and seatbelt use. In one study, people who received normative information about their neighbors' recycling habits, in addition to their own recycling habits, recycled significantly more than those who received information about why they should recycle.

Scaling up social norms

How bar graphs and smiley faces save hundreds of millions of gallons of water each year

At WaterSmart Software, our founders were inspired to use social norms in water management from a 2007 study by California State University (San Marcos) psychology professor P. Wesley Schultz.[9] In the study, researchers distributed leaflets to suburban residents with information about their average energy use and their neighbors' average energy use.

If the household used less energy than the average, researchers provided a smiley face on the leaflet, but if the household used more than the average, researchers drew a sad face. Households that received this bundle of normative information showed significant reductions in energy use following the intervention. More importantly, savings were concentrated among households that initially used above-average amounts of energy; household using less energy than average showed no significant changes in consumption.

WaterSmart Software has scaled up the methodology used in this study to reduce residential customers' water usage across the United

States. Each month, information is sent to consumers with their household's water usage, the usage of similar homes in the neighborhood and the usage of "efficient" homes in the neighborhood.

This information is also paired with a smiley, neutral or worried water droplet icon, depending on the household's water usage relative to other homes; the reports also include personalized tips for reducing water use. Independent analyses state that this information alone helps reduce household consumption rates by an average of 5%.[10]

But even with the right information, water and energy efficiency isn't attributable entirely to the fact that they decrease consumption: these efforts also save money. Efficiency and conservation are often less expensive than other mechanisms used to meet increasing water demand.[11] In addition, some behavioral programs, like those run through WaterSmart, have been shown to significantly increase customer satisfaction and effectively boost the program's goals.[10]

Maximizing the efficacy of social norms

A sophisticated understanding of how a variety of psychological factors interact is needed to help produce meaningful behavior change. The effectiveness of social norms programs may rely on three key factors – injunctive norms, reference groups and data credibility – used in the intervention.

Coupling Descriptive and Injunctive norms

A critically important factor to consider in social norms interventions is descriptive norms versus using injunctive norms to encourage proper water management in a household.

Injunctive norms – which communicate the social acceptability of a norm – are necessary to counteract an undesirable aspect of some social norms interventions, also known as the boomerang effect. This refers to an inadvertent increase in socially undesirable behavior through descriptive norms: by indicating what the typical group *is*,

Ideas for Accelerating Change Around Water | 161

people could begin to waste water because that's what the typical group *ought to do*.

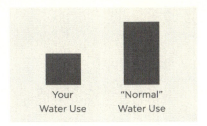

Descriptive norms convey what is typical. In isolation, it may cause individuals who consume less water than the norm to start using more.

Injunctive norms convey what is socially accepted. They may be used to counteract descriptive norms.

Coupling Norms

Injunctive norms, however, can counteract this effect by showing this segment of people that their current behavior is commendable or socially accepted; the smiley faces in Schultz's study is an example of this methodology.[9]

Individual exhibits target behavior

e.g. *low water use*

Message encouraging normalized behavior moves individual back toward the "crowd."

e.g. *"Most people use more water than you."*

An injunctive norm is used to counteract this effect

e.g. *"low water use is admirable"*

The boomerang effect

Water managers should carefully consider the specific visual and written language used to convey these norms. Extensive A/B testing can help to fine tune this language to maximize behavior change.

Reference groups

A second key factor to consider in the design of social norms interventions is reference groups, which are most likely to change an individual's behavior when there is a high degree of perceived similarity or social identification between the individual and the reference group.[12][13]

One way of enhancing an audience's association with a reference group is to provide photographs of appropriate demographic groups in social norms messaging about water use. Ideally, these photos would be of real members who other members of the group might even recognize.[14] Social norms messaging about water use might include photos with local landmarks and people that match a community's demographics.[15]

The exact language used to describe reference groups can play a critical role in determining the success of social norms comparisons. For example, some customers may perceive "neighbors" to literally mean the household next door; using terminology like "similar households in your neighborhood" may be a more effective phrase to increase the likelihood of behavior change.

Equally important as the actual language used to describe reference groups are the analytics used to develop them, such as the number of occupants, the size of a customer's home and local weather patterns. Developing effective reference groups requires not only a large enough sample to allow for a control group and statistically significant results, but also a variety of data sources to ensure a reliable data model.

Data credibility

Data credibility is the third factor critical to consider in social norms interventions. Without credible data, social norms comparisons may fail to achieve intended outcomes.[14] For example, in one study, researchers identified the fact that close to half the students didn't believe social norms data as one of the primary reasons why a social norms experiment to curb alcohol consumption at a U.S. college failed to succeed.[16]

One way of ensuring the credibility of social norms data is to provide comprehensive information about statistics or data cited in social norms messaging, such as when the data was collected and the size of the sample. Another technique is to provide individuals with the opportunity to access more detailed and comprehensive data through a website.[14]

Beyond social norms: other behavioral principles

Social norms are one of the most widely used behavioral tools to reduce resource use. Behavioral science, however, offers myriad other insights that can be applied in water efficiency and conservation programs. Default options, goal setting, public commitments and choice overload are four additional principles that – when applied carefully and systematically – hold the potential to produce further gains in water-use efficiency in the residential sector.

DEFAULT OPTIONS

Psychological and economic research indicates that people are highly motivated to maintain the status quo. Thus, when presented with the choice of sticking with the default option or selecting a different option, people tend to stick with the default option.[17]

Water managers can harness the power of default options by automatically enrolling residents in programs to improve water-use efficiency, and giving them the option to leave the program if they choose. Some research indicates that relatively few people make the effort to opt out. In one instance, just 57 of 20,000 automatically enrolled households chose not to participate in an energy reduction program.

FEEDBACK

Feedback is a common tool used to promote conservation. Like social norms interventions, feedback-based messaging generally provide households with more detailed and timely information about their usage than typical utility bills.[18] Feedback provided to residential energy customers has been shown to reduce energy usage, and combining it with other social science principles – goal setting, commitments and social norms – can increase savings up to 35%.[19,20,21]

In general, research indicates that feedback tends to be more effective when it is provided more frequently; daily or weekly generates the highest savings per household[21] and consumers often value feedback that is delivered frequently. This means that feedback on water resource usage should correlate with the frequency of meter-read data collection.

For example, if water utilities only generate meter reads on a semi-monthly basis, then providing feedback more frequently than that has the potential to undermine data credibility, as discussed above. When more frequent data collection – daily or hourly – is available through Advanced Metering Infrastructure (AMI), these programs can accommodate a more robust feedback schedule, which may lead to greater behavior change.

Water managers interested in feedback-based programs need to take care to manage these effects through segmentation, injunctive norms, and other mechanisms. Feedback is most effective when it is presented clearly, customized to the household, provided over an extended period of time, and when it is provided relative to a meaningful standard of comparison[22]; it's especially effective when it is presented immediately following consumers' actions.[23]

Goal setting

Goal setting can be a powerful tool to increase motivation, overcome procrastination, and set reference points for progress.[24] Setting specific goals can be particularly effective in enhancing effort and persistence.[25]

Goal setting has been used to successfully reduce energy use among residential customers in a number of different studies. In one study of an energy efficiency program employing goal-setting, participating households saved an average of 4.4% in the first year of the program; households that set realistic goals saved substantially more than those who chose the minimum goal possible or set unrealistically high goals.[26] In another study, households assigned a relatively difficult goal of a 20% reduction in energy use saved significantly more energy than households that were assigned a relatively easy goal of a 2% reduction.[27]

Resource managers can utilize the power of goal setting by asking households to set a specific, realistic and achievable (yet difficult) goal. Managers may also consider giving households the option of setting their own goal or having a goal assigned to them. Goal setting can also be used in conjunction with feedback or commitments.

Public commitments

Goal setting is often even more effective when individuals make their goals and commitments public.[28] As a rule, public commitments tend to be more persistent than commitments made privately. Some researchers have suggested that commitment has the power to lead to the development of personal norms that further encourage a behavior, and some evidence suggests that commitments are more effective when people are consistently reminded of them; water managers might consider providing households with written instructions about target behavior to households and asking them to keep them visible.[29]

Commitments to reduce resource use may be especially effective when people make them immediately after being made aware of their high resource use. Participants in one study were reminded of times they had wasted water, then made a commitment to take shorter showers, eventually saved more water than participants in a control condition; they also saved more water than participants who committed to take shorter showers without being reminded when they wasted water.[30]

Water managers may be able to leverage this phenomenon by strategically pairing the opportunity to make commitments with feedback about water usage, as well as leveraging social media to create broader awareness of the public commitment.

CHOICE OVERLOAD

A fifth behavioral principle to consider is a psychological phenomenon known as choice overload, where there's an assumption that more choice will lead to better outcomes.[31] However, recent psychological research indicates quite the opposite: more choices can be paralyzing and could lead to worse outcomes. When people at a gourmet grocery store were presented with 24 types of jam to taste, a study indicates that they were significantly less likely to taste or buy a jam compared to a similar booth with only six types of jam.

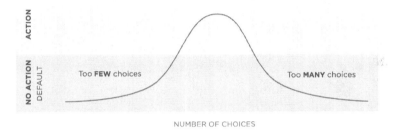

Presenting too many recommendations can overwhelm customers

When applied to resource efficiency, this research suggests that program managers should caution against providing too many conservation and efficiency recommendations, as this can overwhelm customers and make them more likely to stick with the default option: doing nothing.

Resource managers may also consider providing households with packages of efficiency options rather than an item-by-item list of choices to reduce usage.[32][33] While utilities that provide both electricity and water services may consider bundling social-norms based information about both electricity and water, this decision needs to be carefully contemplated, as combined reports risk generating confusion and undermining the potential for behavior change.

PUTTING BEHAVIORAL SCIENCE INTO PRACTICE

Behavioral science offers a range of insights that, when applied appropriately, holds the potential to significantly reduce residential water use. To achieve intended outcomes, behavioral strategies must be applied with a sophisticated understanding of the theory that underlies them and the factors that drive how an insight translates from experiments to the real world.

Water managers should identify the specific outcomes they hope to achieve and carefully select the strategies and insights that will generate these outcomes.[15] They may also choose to employ experimental approaches to evaluate the specific language used to communicate households' consumption patterns, as well as recommendations for reducing resource use. Managers should, at minimum, use statistically valid control and test groups; ideally, more sophisticated testing would generate deeper insights about which elements of a program are effective.[15]

The need for water conservation and efficiency is huge. Americans' water footprint is twice the global average and our available water supply is slowly draining. But, by combining information and financial incentives with well-tested programs that use a range of social and psychological factors, stakeholders have the potential to make the progress necessary to provide reliable access to water for all.

About the Authors

Caroline Hodge - Research Manager, ecoAmerica As ecoAmerica's Research Manager, Caroline is responsible for analysis, write-up, and outreach for a range of social science and marketing research projects that uncover both how Americans think about climate change, and how we can best engage them on the issue.

Jeff Lipton - Director of Marketing, WaterSmart Software. Jeff manages WaterSmart Software's go-to-market strategy development and marketing campaign activities, along with sales operations support.

Endnotes

1   Laitner, J. A. S., Mcdonnell, M. T., & Ehrhardt-Martinez, K. (2013). Consumer Engagement Programs and Smart Energy Efficiency Strategies for Our Nation's Electric Utilities. January.

2   Brown, M. A. (2008). Market failures and barriers as a basis for clean energy policies. Energy Policy, 29(2001), 1197–1207.

3   Geller, E. S. (1981). Evaluating energy conservation programs: Is verbal report enough? Journal of Consumer Research, 8, 331-335.

4   Geller, E. S., Erickson, J. B., & Buttram, B. A. (1983). Attempts to promote residential water conservation with educational, behavioral and engineering strategies. Population and Environment Behavioral and Social Issues, 6, 96-112.

5   Schofield, J. (1975). Effect of norms, public disclosure, and need for approval on volunteering behavior consistent with attitudes. Journal of Personality and Social Psychology, 31, 1126-1133.

6   Sherif, M. (1965). Formation of social norms: The experimental paradigm. In H. Proshansky & B. Seidenberg (Eds.), Basic studies in social psychology (pp. 461-471). New York: Holt, Rinehart & Winston.

7   Sherif, M. (1937). An experimental approach to the study of attitudes. Sociometry, 1, 90-98.

8   J.D. Power & Associates. (2011). 2011 Smart energy consumer behavioral segmentation study℠. Cited in Laitner, J. A. S., Mcdonnell, M. T., & Ehrhardt-Martinez, K. (2013). Consumer Engagement Programs and Smart Energy Efficiency Strategies for Our Nation's Electric Utilities. (January).

9   Schultz, P. W., Nolan, J. M., Cialdini, R. B., Goldstein, N. J., & Griskevicius, V. (2007). The constructive, destructive, and reconstructive power of social norms. Psychological Science, 18(5), 429-434.

10   Mitchell, D. L., & Chesnutt, T. W. (2013). Evaluation of East Bay Municipal Utility District's pilot of WaterSmart Home Water Reports. Report prepared for the California Water Foundation and East Bay Municipal Utility District. (December).

11   Gleick, P. H., Haasz, D., Henges-Jeck, C., Srinivasan, V., Wolff, G., Cushing, K. K., & Mann, A. (2003). Waste not, want not: The potential for urban water conservation in California. Oakland, CA: The Pacific Institute.

12   Burnkrant, R. E., & Cousineau, A. (1975). Informational and normative influence in buyer behavior. Journal of Consumer Research, 2(3), 206-215.

13   Moschis, G. P. (1976). Social comparison and informal group influence. Journal of Marketing Research, 13(3), 237-244.

14   Burchell, K., Rettie, R., & Patel, K. (2013). Marketing social norms: Social marketing and the "social norm approach." Journal of Consumer Behavior, 12, 1–9.

15   McAlaney J, Bewick B, Bauerle J. (2010). Social norms guidebook: a guide to implementing the social norms approach in the UK. West Yorkshire, UK: University of Bradford, University of Leeds, Department of Health. Available at: http://www.normativebeliefs.org.uk/Guidebook.pdf

16   Granfield R. (2002). Can you believe it? Assessing the credibility of a social norms campaign. The Report on Social Norms: Working Paper #2. Little Falls, New Jersey: PaperClip Communications.

17   Samuelson, W. & Zeckhauser, R.J. (1988). Status quo bias in decision making. Journal of Risk and Uncertainty, 1, 7-59.

18   Mahone, A., & Haley, B. (2011). Overview of residential energy feedback and behavior-based energy efficiency. White paper. San Francisco, CA. San Francisco, CA: E3: Energy + Environmental Economics. (February).

19   Fischer, C. (2008) "Feedback on household electricity consumption: a tool for saving energy?" Energy Efficiency, 1, 79-104.

20   Ehrhardt-Martinez, K., Donnelly, K. A., & Laitner, J. A. (2010). Advanced metering initiatives and residential feedback programs: A meta-review for household electricity-saving opportunities. American Council for an Energy Efficient Economy (ACEEE) Report No. E015. Washington, D.C.: ACEEE. Available at: http://www.aceee.org/research-report/e105

21   Darby, S. (2006). The effectiveness of feedback on energy consumption. Working Paper. Oxford, UK: Oxford Environmental Change Institute. (April).

22   EPRI (2009). Residential electricity use feedback: A research synthesis and economic framework. EPRI Report No. 1016844. Palo Alto, CA: EPRI. (February).

23   Seligman, C., & Darley, J. M. (1977). Feedback as a means of decreasing residential energy consumption. Journal of Applied Psychology, 62(4), 363–368.

24   Todd, A., & S. Houde. (2010). List of behavioral economics principles that can inform energy policy. Available at: http://www.annikatodd.com/List_of_Behavioral_Economics_for_Energy_Programs.pdf

25   Locke, E. A., & Latham, G. P. (2002). Building a practically useful theory of goal setting and task motivation: A 35-year odyssey. American Psychologist, 57(9), 705-717.

26   Harding, M., & Hsiaw, A. (2012). Goal setting and energy efficiency. White paper. Available at: http://www.stanford.edu/~mch/resources/Harding_Goals.pdf

27   Becker, L. J. (1978). Joint effect of feedback and goal setting on performance: A field study of residential energy conservation. Journal of Applied Psychology, 63(4), 428-433.

28   Abrahamse, W., Steg, L., Vlek, C., & Rothengatter, T. (2005). A review of intervention studies aimed at household energy conservation. Journal of Environmental Psychology, 25(3), 273–291.

29   Cialdini R.B. & Trost M.R. (1998). Social influence: social norms, conformity, and compliance. In D.T. Gilbert, S.T. Fiske, & G. Lindzey (Eds.), The handbook of social psychology (4th ed.) (pp. 151–92). Boston: McGraw-Hill.

30   Dickerson, C. A., Thibodeau, R., Aronson, E., & Miller, D. (1992). Using cognitive dissonance to encourage water conservation. Journal of Applied Social Psychology, 22, 841-854.

31   Iyengar, S. S. & Lepper, M. (2000). When choice is demotivating: Can one desire too much of a good thing? Journal of Personality and Social Psychology, 76, 995-1006.

32   Opinion Dynamics Corporation & Navigant Consulting (2012). Massachusetts three year cross-cutting behavioral program evaluation integrated report. Prepared for Massachusetts Energy Efficiency Advisory Council & Behavioral Research Team, p. 3.

33   Vörösmarty, C.J., Green, P., Salisbury, J., & Lammers, R.B. (2000). Global water resources: vulnerability from climate change and population growth. Science, 289, 284–288.

[20]

Conserve2Enhance: Bridging voluntary water conservation and community-driven environmental enhancement – Candice Rupprecht, Dr. Sharon B. Megdal, Brittany Xiu and Kelly Mott Lacroix

Putting a price on "the right thing to do"

What happens to the water that I conserve or where does the saved water go?

In growing areas where water is scarce relative to demands, water conservation behaviors often mean existing water supplies are channeled to support increased demands of municipal, commercial and industrial water users. In many cases throughout the semi-arid

Southwest, river systems and natural areas have been degraded as water supplies have been directed to meet human demands.

Although most recognize that conserving water is "the right thing to do", some have questioned why they should conserve water if the saved resource is later directed to growth. It was this very question that motivated the University of Arizona Water Resources Research Center (WRRC) to develop the Conserve2Enhance™ program.

Known as C2E, this program connects water conservation with the desire to enhance natural systems, and it demonstrates how collective community action can lead to meaningful environmental benefits. C2E offers a conservation solution that gives people an opportunity to decide exactly where their water savings are going.

The concept is simple: participants implement conservation measures in their home or business and keep track of their water use. Using a mechanism developed specifically for the purpose of associating a monetary value to the saved water, such as the C2E Water Use Dashboard, ratepayers can keep track of their home usage and have the ability to voluntary donate to a fund for environmental enhancement projects. By placing a monetary value to the water savings, participants are able to see how those savings directly benefit environmental projects of value to local communities or organizations.

When the WRRC unveiled Conserve2Enhance™ in 2008, it was described as an innovative mechanism to raise funds to purchase water for environmental restoration projects through municipal water conservation. As the program has evolved, so has the focus of project funding.

From purchasing water for restoration projects to including temporary watering, plants, and materials as well as design fees and site signage, a key aspect of the program is the ability to support community conservation goals by directly linking residential and commercial water conservation actions with environmental enhancement efforts.

Project types may include riparian restoration, urban wildlife enhancement, green infrastructure installation, in-stream flow protection, erosion control, flood mitigation and source water protection.

By providing a much needed, community-driven source of revenue to help implement already identified projects, funding projects like these can also motivate communities to identify environmental priorities and establish a process for selecting projects.

Key elements to this community-driven program

While the concept is simple, the reality of implementing a program like C2E can be quite complex. As our first program was being developed in Tucson, Arizona, and the C2E mechanism was promoted across the Colorado River Basin, three essential elements for a successful program became evident:

1. the program required support from local stakeholders,
2. enhancement project(s) or priorities needs to be approved by the community itself, and
3. an accounting mechanism to track participants' water use and savings.

Whether the program is developed by a utility or a NGO (non-governmental organization), buy-in for the program concept and tailoring the program to the region are extremely important. A community-driven program needs to be locally supported by organizations with the capacity to manage both participants and projects.

Innovative programs, such as C2E, can take time to build traction and require sustained effort to grow. Although many people have embraced the concept, getting communities and organizations to brainstorm how the program could work has proved challenging, even with the tools that exist to support program development.

However, once a core group of stakeholders is established, the identification of environmental priorities is necessary. In some communities, this type of prioritization may have already been completed and this element will be as simple as referring to a pre-established technical or guiding policy document. In other communities, the path forward may be less clear; more information, later in this chapter, will explain how Tucson's C2E pilot program is a prime example.

THE TOOLS FOR CONSERVATION-BASED SUCCESS

The Conserve2Enhance™ brand features a trademarked name and logo, along with tools to help guide development and implementation of both local and regional programs. The three primary tools are the *Program Design Guide*, the electronic *Program Development Toolkit* and the Water Use Dashboard.

The *Program Design Guide* features a comprehensive, 61-page document that presents various options for the development of the program, which also includes a readiness rating worksheet; this helps communities and organizations determine whether a C2E-type mechanism is appropriate for their environmental goals. The electronic *Toolkit* is a compilation of all the resources that have been developed to administer and promote the program, and all toolkit materials are customizable to meet the needs of a community.

Launched in 2014, the Water Use Dashboard educates the public about their water resources, provides management tools to track water savings, and facilitates program donations (Figure 1). By creating a customizable tool to house participant water-use data, educational resources, water conservation messaging and participant donations, the Dashboard eases the adoption of C2E for new communities, organizations and businesses, and it opens the door to multi-scale water conservation-based donation programs:

Dashboard Data: Once data is uploaded to the site, participant water use information is displayed as a breakdown of indoor and outdoor water use, average annual and monthly water use, daily water savings and comparison of individual water use with their community's average water use. As seen in Figure 1, this information is presented through simple graphs and an easy-to-read water use "box score".

Educational Resources: Emphasizing the connection between individual water use and the environment, consumers are provided with actionable information, such as links to water conservation resources, information about how individuals use water, and being able to track water from its source to its users.

Customized Messaging: In addition to a section on indoor and outdoor water conservation options, program managers and utilities can connect their existing conservation and rebate programs to the mechanism, which provides additional motivation and an opportunity for participants to donate funds. Group email messages can also be sent to participants based on website activity, geographic area or affiliation.

Donation Platform: A password-protected donation portal has been established using PayPal that displays suggested donation amounts, past donation history and allows them to set up recurring donations. Suggested donations are based on a participant's monthly water use and savings, ranging from $1 to $10. Donations can be made to local and regional programs, depending on the participant's preference.

Figure 1 - Screenshot images of the C2E website: www.conserve2enhance.org. Left: C2E homepage featuring navigation and call to "Join C2E"; Right: C2E Water Use Dashboard featuring a sample participant box score, scrolling graphs of water use, donation summary and a seasonally appropriate conservation tip.

C2E at work: the Tucson pilot program

WRRC's concept was first unveiled in the *Journal of the American Water Works Association* in a 2008 article titled "Conserve to Enhance – Voluntary Municipal Water Conservation to Support Environmental Restoration," which was authored by Dr. Sharon B. Megdal and Andrew Schwarz. Later that year, a core team of members from Tucson Water, the WRRC and two local NGOs (the Sonoran Institute and the Watershed Management Group) turned that idea into a pilot program.

In 2009, an advisory board was formed to ensure local values were represented in the program design and to establish oversight for the management of its funding. Current and former advisory board members include conservation organizations, local officials and business representatives. Throughout 2010, the board also worked to develop criteria for selecting a riparian restoration site to invest funds, and they create a way for Tucson Water ratepayers to donate to the cause directly from their water bills.

In 2011, the Tucson C2E pilot program launched with 60 participants, 29 of whom installed rainwater harvesting projects at their homes through subsidies from a U.S. EPA grant. From January 2011 through December 2013, pilot participants saved over 3.2 million gallons of water, and nearly $37,000 was donated by Tucson Water customers and project participants.

The advisory board also awarded $12,000 to the Tucson Audubon Society to replant native vegetation lost due to drought and floods, and they were able to create a demonstration site for passive water harvesting techniques in Atturbury Wash (Figure 2). The site was a perfect for the first project because it is adjacent to an ephemeral wash within a large regional park that provides high visibility.

Figure 2 - Volunteers working to dig water harvesting basins and create a raised trail system with the excavated soil at Atturbury Wash in 2012.

The Tucson C2E Community Enhancement Project grant program was formalized in 2013 and now serves as an annual opportunity to invest program funds in the Tucson community; funding was awarded to three community projects in 2013 and three more in 2014. These six project sites – including an elementary school, several neighborhoods and a city park – will help mitigate stormwater flooding and remove invasive species in addition to enhancing urban waterways and wildlife habitats in the Tucson urban corridor. In total, nearly $60,000 has been invested in local environmental projects within this short period of time.

COLLECTIVE DROPS IN THE BUCKET OF SOLUTIONS

Community-driven environmental enhancement projects offer a creative solution for addressing the challenges faced by our river systems, riparian areas and urban corridors. Individuals and communities can invest their conservation savings in meaningful projects, which will benefit their local water systems and adjacent habitats.

Providing opportunities for individuals to get involved is one important step in building the momentum to address these challenges. Voluntary conservation alone will not solve regional water challenges, but bridging municipal water conservation and environmental enhancement is one way for water conservation to support other environmental-friendly projects, whether it's across town or across the state border.

About the Authors

For the past seven years Candice Rupprecht has worked on numerous projects at Water Resources Research Center, developing innovative solutions for water management and water resources education. The hallmark of these projects is applied research coupled with focused stakeholder engagement. She holds a B.S. in Geosciences and a M.S. in Hydrology, both from the University of Arizona. As of September 2014, she will begin working as a conservation specialist in the Strategic Initiatives Division at Tucson Water.

Candice Rupprecht - candicer@cals.arizona.edu
Sharon B. Megdal - smegdal@cals.arizona.edu
Brittany Xiu - brittanyx@email.arizona.edu
Kelly Mott Lacroix - kmlacroix@email.arizona.edu

Water Resources Research Center
The University of Arizona
350 N Campbell Ave
Tucson, AZ 85719

[21]

Rainfrastructure: Building stormwater utilities – Kevin Mercer

Our challenge: no pipe big enough

"Where are we with stormwater?"

That rarely resonates as the first question urban planners, ecologists and insurers ask themselves. Here are three reasons why these professionals may soon want to revisit this question:

1: Stormwater runoff is becoming very costly.

According to the Insurance Bureau of Canada, stormwater runoff – and its big brother, overland flooding – is already one of the largest sources of property and casualty claims in North America, frequently larger than fire and theft combined.

Consider the damages from Superstorm Sandy and the unnamed floods of 2013 and 2014 that overwhelmed Calgary, Minneapolis, Detroit, Toronto and Boston. Other locations have had major (and "minor") storm and flood damage, which have annually topped *billions* in damage claims and threatened the livelihood of tens of thousands.

2: The runoff is a major threat to ecosystem and potable water security

Generally unknown to the public, stormwater runoff is the largest source of freshwater pollution and one of top threats to drinking water security.

Actually, let me correct myself because I may have erred in my estimation of the public's perception of that risk... at the time of this book's publication, I'm sure the residents of Toledo, Ohio are fairly cognizant how stormwater runoff impacts our drinking water infrastructure and people's livelihoods.

3: Rainwater is beginning to take its toll on our sewer systems.

Climate change has altered precipitation patterns and our "normal rainfall" is now the single largest threat to the fiscal security and sustainability of our cities. Existing urban sewer infrastructures are thoroughly inadequate, and rebuilding it is an unfunded liability that measures in hundreds of billions of dollars.

The typical hot, fast and dirty rampage of "normal" rain is a major concern because our cities cannot afford the cost, or the social disruption, to completely rebuild conventional infrastructures that would accommodate this new climate change reality.

Low-impact development can relieve some of these issues

Conventional infrastructure methods have not proven sufficient, so utilities are turning to low impact development (LID) methods to control their ballooning infrastructure costs. But, in doing so, they come face to face with a whole new set of operational and social challenges associated with managing stormwater.

So far, the most dominant method of coping with urban stormwater is to establish utilities that are focused specifically on stormwater usage. Although there is nothing abnormal about establishing municipal utilities to cope with emerging infrastructure needs, these utilities somehow come with a whole new agenda of extensive economic, health, social, and environmental issues. These concerns will challenge our expectations and understandings around the role of infrastructure in our cities.

Internationally, the trend is toward to offset the costs to mitigate the negative impacts of changing precipitation patterns and increased urbanization. To do so, over 1,400 North American jurisdictions have implemented stormwater utilities, often with rates correlated to property size or impervious cover, and hence volume of runoff. While this realignment of the traditional property owner subsidy is often decried as a new tax, it is anything but.

How come? By charging for stormwater, communities are actually removing an unfair subsidy received by property owners who discharge unlimited volumes of polluted runoff, especially when the local system is unable to cope with it; in cases where those discharges result in combined sewer overflows, the property owner is associatively responsible for their downstream impact as much as that of the utility.

But utilities and their fees are not effectively countering the financial and operational challenge of stormwater. This particular water sector has widely adopted best management practices (BMPs) for technical and institutional procedures to store and treat stormwater. Additionally, many permits and consent decrees in the US now mandate substantial green infrastructure expenditures rather than simply expanding the present-day methods.

LID practices have gained momentum, primarily focusing on handling rainwater at or near where it falls, but can these systems actually provide effective stormwater management with lower capital and comparable operating costs? Can communities surmount the challenge of social engagement necessary to install and maintain these facilities on private property?

Innovation Means Smart Change, Not Big Change

One of the greatest challenges our cities face is the impact of climate variable stormwater. It can increase pollutants that threaten the key sources of drinking water, and unregulated runoff and overland flooding on our urban infrastructures can challenge the sustainability of those infrastructures.

There are also concerns with formal municipal or stormwater utility managed assets, in addition to the conversion of do-it-yourself residential projects that manage rainfall runoff. The typical poster child for this type of initiative is the prosaic residential rain barrel.

Nothing illustrates the yawning chasm of differentiation between taking rain seriously, and our current indifference to its importance in our lives (and urban infrastructures), more than the practice of harvesting rain. One of the most promising innovations in property-based LID systems in today's stormwater market is the concept and application of raingrid residential cistern networks.

Raingrid networks automate traditional stormwater management technology by fitting rain barrels/cisterns with real-time sensors and the ability to remotely manage their operation. In doing so, it is conceivable that a stormwater smart grid can be installed at an estimated cost of $1,500 per property, $60,000 per hectare, or $100 per cubic meter of storage; this is well below half the cost of conventional stormwater storage methods (excluding land costs).

A raingrid setup utilizes a smart controller and data management system to provide remote, real-time cistern operations coupled to a central server that calculates storage volumes against predictive precipitation. Municipalities will be able to create a smart grid that captures and stores rainfall *before* it becomes stormwater.

The 5-day predictive precipitation algorithms of the central server calculate available storage against need and remotely release stored water in the event greater storage is needed to cope with predicted rainfall. In doing so, the cisterns create reliable, sufficient storage capacity to eliminate (or reduce) overland flooding – and minimize property damage. This therefore allows utility planners to reduce or eliminate the need to super-size pipes or retention pond networks in order to cope with climate variable stormwater impacts on a city.

In doing so, and most significantly, a raingrid network transforms distributed stormwater methods from what it has heretofore been, a dispirit collection of passive, non-standardized problems, into a reliable, smart, managed stormwater asset.

A LEVEL PLAYING FIELD

To compete on a level playing field with conventional system expansion or end-of-pipe technology, point-of-source technologies must be rolled out on a neighbourhood or citywide utility scale. Even with fairly inexpensive cisterns and intelligent controls, a rainwater network system is still a multi-million dollar purchase, and the prospect of spending millions for a system installed on private properties is still anathema to most municipal asset managers.

This policy infancy of stormwater utilities, coupled with the typically long sales cycle inherent of municipal purchasing, has consequently hindered acceptance and growth of smart distributed stormwater technologies. Even municipalities with extensive rainbarrel programs, such as the program in Washington, D.C., face a long path to system acceptance and integration from an engineering community that's just not convinced that they *want* to work on private property.

Municipal acceptance of this technology, and the willingness to house municipal assets on private property, is a new business model that adds further complexity to the sales cycle and financing of the system. Networked residential stormwater cisterns may be more effective and efficient from an infrastructure operations and financing standpoint, but, if their rate of adoption is slow, it may prove difficult to secure project financing. It can be difficult for utilities to introduce not only a new technology but also a new business model.

How Utilities Can Work This Out

By replacing conventional stormwater methods with point-of-source solutions, our company, RainGrid, has been able to blur the technology adoption barrier dividing the traditional home and state division of infrastructure.

The method we have devised to cope with this unconventional disconnect between public and private property, is by means of developing a truly unique public-private partnership between private property owners and their stormwater utilities. Agencies – instead of property owners – will purchase, install, and maintain the system's cisterns as a municipal asset.

The installation of publicly or utility-owned and managed infrastructure on private property is not a new concept; this is similar to gas lines, electrical smart meters, and sewer laterals. However, what is unique about this stormwater system arrangement is that both the homeowner and municipality reap extraordinary benefits.

In exchange for hosting no-cost network system installations, property owners receive reduced stormwater utility rates, plus access to stored rainfall to offset potable water demand for irrigation purposes or internal building purposes. In that regard raingrid distributed stormwater infrastructure also provides a cross-beneficial supply of water to reduce potable water infrastructure capital and operating requirements.

Transform Stormwater into a Resource

Despite stormwater damage being one of the largest sources of insurance claims, and the emerging acceptance of distributed stormwater management as a viable and even legally mandated solution, there remains a noteworthy disconnect between landowners' behavior and municipal responsibility.

The general public's lack of knowledge concerning water management challenges, as well as its limited understanding of innovative options available to mitigate disruptive flooding, is but one side of the primary hindrance of our ability to finance the widespread deployment of new technologies that includes the community as part of the solution. An equally limited lack of knowledge by stormwater utilities and political decision-makers of the innovations available to resolve these challenges with distributed technologies instead of relying upon the conventional means, stands as the other.

Continued climate disruption, especially extreme weather events, will rapidly raise public demand for economical environmental solutions. This increased awareness and presence of innovative public-private partnerships will start to benefit the acceptance and success of the raingrid infrastructure technology for stormwater utilities and their customers.

Additionally, the market for this technology is extensive, and it can support multiple companies; this healthy competition will help break down barriers and will accelerate the adoption of – as we affectionately call it – the rainfrastructure.

Big Data from Distributed Sources

While the initial and most valuable outcome of building a smart raingrid technology lies in the rainfrastructure itself, there is a concomitant value to the very reason we pursue new frontiers of climate adaptation innovation. Every raingrid system developed and installed will generate a mammoth volume of micro-climate data – and the data is so powerful that it will be possible to draw a lot-by-lot assessment of our cities' stormwater risk.

Given that wealth of data, remember the three challenges posed by stormwater: cost, security and adaptation. The granular micro-climate data can turn into cost-effective information that empowers urban planners, infrastructure designers and builders; it will enhance the efficiency of methods used by property builders and owners; and the insurance, mortgage and related service providers can underwrite the financial demands of our cities with a fair level of certainty to keep our cities sustainable.

Now that's water innovation worth pursuing.

About the Authors

RainGrid co-founder and CEO, Kevin Mercer designs municipal rain barrel programs, a social marketing specialist and manufacturer of the RiverSafe RainBarrel.

RainGrid co-founder and CTO, Stephen Braun is a registered engineer specializing in stormwater filter design and a hydraulic modelling specialist.

Jennifer Stoneburgh is an associate with the MaRS Discovery District in Toronto in cleantech, physical sciences, and advanced materials. All three are located in Toronto Canada.

[22]

From Serious Issues to Serious Games - and Back! – Chengzi Chew & Gareth James Lloyd

WHEN FACED WITH DRY TOPICS OR COMPLEX ISSUES, many students have been stimulated by activities that they consider to be informal and fun; in other words, playing a game makes it easier to learn.

The same idea should work for raising awareness on the scarcity of water, providing insight on new ways to recycle our resources and educating the public about a topic that becomes important when the taps start to run dry. A desirable approach is one where the fun activity part is combined with more formal activities, which should increase the potential for learning.

Such an approach is applied through serious gaming, and it is being used with success in many areas. From health care to city planning, from engineering to military tactics, games are being used to enhance and expand upon the educational experience. That is why we are advocates of using games to raise awareness and educate in the field of water.

THE VIRTUAL BLOCKAGE WITH WATER AND DATA

There are critical links between social and economic developments, as well as environmental sustainability, where water is an essential building block. However, these connections are not easily recognized, which leaves room for ineffective or poor water-planning decisions.

Without a full understanding how water should be used, there's indecision whether the water should be allocated between various domestic, agricultural, industrial and/or energy users. This is often the result of silo-like thinking.

For example, when an upstream hydroelectric power plant underestimates (or simply does not consider) the need to consult and coordinate releases of water with downstream agricultural users, town planners and smaller communities, issues will arise in the future. And, when developing countries grow past a certain point, there will be a need for holistic planning approaches and practices when it comes to water.

The first step in improving the development, management and use of water is to raise the awareness on how users and uses are very closely interconnected – even if it doesn't initially appear to be connected in any way. Various information campaigns and training programmes have been finding new ways to make these efforts more efficient, more effective and scalable in order to achieve the desired progress. A major opportunity to accelerate progress is through the power of information, where we utilize the advances in and dissemination of information & communications technology (ICT).

One of the main upsides of digital communication and information is accessibility, where people can communicate socially, organize financial transactions and keep up to date with the latest local or international news. However, one of the main downsides of digital communication is the overabundance of information, whether it's a flood of Twitter updates or a barrage of advertisements through e-mail. We are exposed to these sources at such a high rate that it affects the way our brains process information (Small G. and Vorgan G., 2008 and Gee JP, 2003).

Daily exposure to high technology – computers, smart phones and video games – stimulates brain cell alteration and neurotransmitter release, gradually strengthening new neural pathways in our brains while weakening older ones. Today, every single source of information (from books to text messages) competes for our attention; the most interactive and engaging content typically wins.

Building Awareness in a World of Distractions

We now face additional challenges in capacity building and raising awareness, besides having to make efforts more efficient and scalable. We also need to keep up with communication trends (on new pieces of technology) to be able to attract the attention and interest of today's audience.

If we think about grade school, then we can accept how students are stimulated by activities that they consider to be informal and fun. (Nearly everyone would like going out for recess.) But students aren't as stimulated by more formal activities and associated materials that schools consider of greatest importance (after Stapleton, A J, 2003; Shute, V. J., Ventura, M., Bauer, M. I., & Zapata-Rivera, D., 2009). This is where serious games can bridge the informational gap.

Games that are used for purposes other than entertainment are commonly defined as serious games; it's not a new concept, but is growing in acceptance. Serious games using real-life data – coupled with real-time modeling plus social and economic factors – opens up a new paradigm. For active stakeholder participation and education, this is where we think games can help inspire serious change.

Why Games Are Great

We believe that games are excellent at getting people to be interested in topics and getting people to learn things at a faster rate; games even help with memory skills.

Confucius once said, "I hear and I forget. I see and I remember. I do and I understand" - we can use this piece of philosophy to study the way we learn how to ride a bicycle. Someone gets on to the bicycle, tries to balance and the bicycle reacts. When we fail at first, our body adjusts and learns to balance a little differently. This cycle continues until we finally learn the correct way to do it.

The learning experience is similar in a well-designed game environment, where it provides a feedback mechanism so players can reflect on their actions and adopt a different approach to win the game. This internalization of actions and reactions is what referred to as "meaningful play", a condition described by Katie Salen and Eric Zimmerman in "Rules of Play". Games also shift the focus of control in learning from the teacher to the player, so the environment enforces self-learning techniques and impacts knowledge retention.

Serious games should have as much meaningful play as possible, so content in the game should also be as realistic as possible. A serious game should help players to achieve realistic goals; this is a major reason why video games are now being used to help soldiers recover from PTSD.

When we design a game, our ideas on realism are currently limited to the content, rather than the graphics. Sure, we would love to make blockbuster serious games with really cool graphics, but that requires a budget of a Hollywood blockbuster. However, that doesn't stop us from making a great game.

What are we doing?

We've been developing a game called Aqua Republica that can be modified to produce suit different applications. Aqua Republica combines a scientific model (MIKE HYDRO) with a game engine, and the two systems interact with each other while someone is playing the game.

The scientific model is used by various governmental groups to help calculate and forecast water flows in river basins, and that data is analysed to make management decisions. To understand the physical environment and interactions within a river basin, we think there's no better way to learn than to the use realistic data, combined with the laws of physics and hydrology, as a portion of the game rules.

The game engine uses the results from the model and is linked to other parameters, such as population, economy and other social issues related to water. Since nothing remains constant in real life, the player interacts with the scientific model to produce evolving scenarios that players must react to. Adapting and improvising becomes an important part of the game, so the interaction between the scientific model and the player provides an interesting (and very relevant) learning experience.

These learning experiences can also be used as a way of sharing and discussing scenarios based on data and model results with non-expert stakeholders. For example, imagine a model using climate change data and local water data, then that information is linked to the local water issues in a game. When someone plays the game, he or she also gets to learn about details of local water issues.

By creating interest through gaming, players may become more aware and curious about these issues, which could inspire them to learn more about them. This could create a sequence where people will start to come up with ideas to solve these issues. An example of this approach is an ongoing research project in the Middle Oliphants Catchment in South Africa, where Aqua Republica is being used to support stakeholder negotiation processes.

THE WAYS AHEAD

Thirty years ago, people thought games were a waste of time, where players would become couch potatoes and not contribute to society. Look at how times have changed.

> We believe that there are multiple ways ahead for serious games, and that we are only scratching the surface of how we can use them to affect development, education and awareness of water issues.

With Aqua Republica, we are focusing on educational gaming competitions, which are great ways to get students interested in water and encourage schools to spark innovation in the classrooms. For example, our annual UNEP-DHI Eco Challenge school competition is growing tremendously from just 30 teams in 2013 to over 400 teams in 2014. We are currently considering options to bring the learning experience to more schools around the world.

There are many more possible ways to advance this initiative, such as further enhancing of the interactions between game engines and scientific models so we could harvest the players' results in the game to be used as solutions. This idea is developed from the puzzle game called "Foldit", where players fold proteins in the game and the highest-scoring players would be recognized by a researcher.

In 2011, players of Foldit helped to decipher the crystal structure of the Mason-Pfizer monkey virus (M-PMV) retroviral protease, an AIDS-causing monkey virus. While the puzzle was available to play for a period of three weeks, players produced an accurate 3-D model of the enzyme in only ten days. This is a phenomenal accomplishment because the structure of the enzyme had been a mystery for scientists for nearly 15 years.

The same idea could be made in Aqua Republica, where the player with the highest score could result in contact with a planning researcher. And this could lead to brainstorming activities, such as the most efficient land-use and water resource planning in river basins or how to crowdsource solutions from local stakeholders.

We also think that educational accreditation for such games would pave the way forward. Serious games are a great supplement to existing classroom materials, and getting a stamp of approval by education boards would directly impact students all over the world; if an approved game can help a student learn faster and retain that knowledge, then we've done our job.

Finally, remember that games cannot be limited to high school students. According to the Entertainment Software Association (ESA), the average age of a gamer is 31 years old and 48% of gamers are female. With more in-depth content that involve different themes, serious games can benefit a larger number of professionals and continue as part of a lifelong learning process.

ADDITIONAL INFORMATION

Aqua Republica game website – www.aquarepublica.com
UNEP-DHI Eco Challenge website –
http://www.the-eco-challenge.org/

About Chengzi and Gareth

Chengzi leads the serious gaming initiative at DHI. He started the Aqua Republica game platform in 2013 together with Gareth. Chengzi has a background in hydro-informatics and water management and has previously worked with numerical modelling and programming. He also designs and plays games. Follow him on twitter @chengzichew.

When not playing serious games with Chengzi, Gareth uses a fair portion of his time working as Senior Advisor at the UNEP-DHI Centre, where he applies his knowledge of water and water-related issues to support the United Nations. Gareth has a background in environmental and development policy and has worked in journalism, education and in software development for Microsoft.

Endnotes

Gee JP (2003a), High Score Education: Games, not school, are teaching kids to think, http://www.wired.com/wired/archive/11.05/view.html?pg=1. Accessed 5 Jul 2013

Gee JP (2003), What Video Games Have To Teach Us About Learning And Literacy, Palgrave Macmillan, New York, NY

Stapleton AJ & Taylor PC (2003), Why Videogames are Cool & School Sucks! Paper presented at the annual Australian Game Developers Conference, Melbourne, 20-23 Nov2003

Stapleton AJ & Taylor PC (2002), 'Physics and Playstation Too: Learning Physics with Computer Games' Australian Institute of Physics 15th Biennial Congress, Darling Harbour, NSW, 8-11 July 2002

Stapleton A (2004) Serious Games: Serious Opportunities. Paper presented at the Australian Game Developers' Conference, Academic Summit, Melbourne, 2 Dec 2004

Small G and Vorgan G (2008), iBrain: Surviving the technological alteration of the modern mind. Collins Living, Harper Collins Publishers

Shute VJ, Ventura M, Bauer MI, & Zapata-Rivera D (2009), Melding the Power of Serious Games and Embedded Assessment to Monitor and Foster Learning, Chapter 18, Serious Games: Mechanisms and Effects, p 295-321

Salen K and Zimmermn E (2004), Rules of Play: Game Design Fundamentals. Cam-bridge, Mass., and London, England: The MIT Press

Game Player Data, http://www.theesa.com/facts/gameplayer.asp Accessed 1st April 2014

Acknowledgements

Art/Design

Thanks to Angelina Gonzales of N3Q Media (http://www.nthreeq.com/) for providing the design and artwork for the cover.

Content Editor/Proofreaders

Thanks to Craig Martin (http://craigmartinbusinesswriter.com) for his constant flow of editorial advice – without which this book would never have been complete.

Family

I thank the love of my life, Truc-Chi, who helped bring this project to bear with her love and calming presence during times of doubt and insanity. With undying love and devotion, I dedicate this project and all I do to help design a better future to my two incredible sons, Ethan and Adam.

Readers

For everyone who has read this anthology, for those who have reviewed this work, and for all of the fans of the amazing individuals who contributed chapters, thank you for your support! Without you, the readers, this book would never have been possible.

Writers

And a huge thank you to all of the authors who appear in this anthology. It was a privilege to work with each one of you on this project. I look forward to our ongoing relationships and your future innovative ideas!

About the Editor

CHRISTOPHER J. PEACOCK has been called an Entrepreneurial Water Maverick by some. He is the founder of The Water Innovation Project and is on a mission to transform the way we value water though innovation and collaboration. He is also the founder of H2.O, an online collaboration platform for the water industry to share information and knowledge around water and data. During the day, Chris works with FATHOM™ leading Strategic Accounts and helping to bring their revolutionary SaaS, cloud based platform to Water Utilities across North America.

Chris is always on the lookout for amazing people and exploring new ideas to manage water. He can be reached through www.waterinnovationproject.com or via twitter @aeleader.

Copyright Acknowledgements

Damned If We Don't!
Ideas for accelerating change around water
© 2014 by Christopher J. Peacock

Introduction
© 2014 by Christopher J. Peacock

"Extreme Weather in Canada - Math & Aftermath"
© 2014 by Robert Sandford

"Integration and Consciousness"
© 2014 by Suzanne King

"The End of Traditional Conferences: How water discussions must evolve"
© 2014 Karen Kun

"Developing Leadership in the Water Sector"
© 2014 Renee Kayal

"When Women's Networking Gets Personal (in the good way)"
© 2014 Leila Munla

"Water Issues are People Issues"
© 2014 Julia Ko

"Advancing One Water Management with One Water Communication"
© 2014 Donna Vincent Roa, PhD, ABC

"Utilize Social Media – don't block it."
© 2014 Todd Danielson

"Getting beyond dots on a map: Collaborative science and scenario planning in the Upper Gila River Watershed"
© 2014 Kelly Mott Lacroix & Dr. Sharon Megdal

"Water Clusters and the Acceleration of Innovation"
© 2014 Jon Grant

"Accelerating Innovation"
© 2014 McGee Young

"Technology, Utilities and Innovation - from "What if?" to "Here's how""
© 2014 Jim Ginley

"Smart Homes, Smarter Consumers"
© 2014 Eric Elias

"Why We Need to Unlock the Enormous Power of Data in the Water Industry"
©2014 Sunit Mohindroo & Ahmed Badruddin

"Industrial Ecosystems: 21st Century Challenges, 21st Century Solutions"
© 2014 Jim Lauria

"Convergence: Utilities need to mind the data as well as the drops"
© 2014 Graham Symmonds

"The Hitchhiker's Guide to Digital Efficiency"
© 2014 Alan Hinchman

"Non-Revenue Water: A prime target of water innovation"
© 2014 David McGimpsey

"Tapping into the Power of Behavioral Science: Insights & opportunities for water-use efficiency"
© 2014 Jeff Lipton & Caroline Hodge

"Conserve2Enhance: Bridging voluntary water conservation and community-driven environmental enhancement"
© 2014 Candice Rupprecht, Dr. Sharon B. Megdal, Brittany Xiu and Kelly Mott Lacroix

"Rainfrastructure: Building rain grid stormwater utilities"
© 2014 Kevin Mercer

"From Serious Issues to Serious Games - and Back!"
© 2014 Chengzi Chew & Gareth James Lloyd

Cover Art by Angelina Gonzalez

Continue the Discussion

WE HOPE YOU'VE ENJOYED READING *Damned If We Don't! Ideas for accelerating change around water.* If so, we invite you conspire with us, push the boundaries and help us accelerate change.

Visit www.wateranthology.com to learn more about this book and see full color images. You can also sign up for our newsletter to read interviews with authors, stay informed about special announcements, receive invitations for events and hear about upcoming book releases.

Learn more about Water and Data by joining H2.O – the go to place for everything related to the smart water grid and water related data! www.H2dotO.com

Follow us on Twitter
@H2OInnovate

Befriend us on Facebook:
https://www.facebook.com/wateranthology

Made in the USA
San Bernardino, CA
13 October 2014